Lecture Notes in Statistics 129

Edited by P. Bickel, P. Diggle, S. Fienberg, K. Krickeberg,
I. Olkin, N. Wermuth, S. Zeger

LIBRARY

U

Springer
New York
Berlin
Heidelberg
Barcelona
Hong Kong
London
Milan
Paris
Singapore
Tokyo

Wolfgang Härdle
Gerard Kerkyacharian
Dominique Picard
Alexander Tsybakov

Wavelets, Approximation, and Statistical Applications

 Springer

W. Härdle
Humboldt-Universität zu Berlin
Wirtschaftswissenschaftliche Fakultät
Institut für Statistik und Ökonometrie
Spandauer Strasse 1
D 10178 Berlin

G. Kerkyacharian
Université Paris X
URA CNRS 1321 Modal
200, av. De la République
92001 Nanterre Cedex
France

D. Picard
Université Paris VII
UFR Mathématique
URA CNRS 1321
2, Place Jussieu
F 75252 Paris cedex 5
France

A. B. Tsybakov
Université Paris VI
Laboratoire de Probabilités
B.P. 188, 4, Place Jussieu
F 75252 Paris cedex 5
France

Library of Congress Cataloging-in-Publication Data
Wavelets, approximation, and statistical applications / Wolfgang
  Härdle ... [et al.].
      p.   cm. -- (Lecture notes in statistics ; 129)
  Includes bibliographical references and indexes.
  ISBN 0-387-98453-4 (softcover : alk. paper)
  1. Wavelets (Mathematics) 2. Approximation theory.
3. Nonparametric statistics. I. Härdle, Wolfgang. II. Series:
Lecture notes in statistics (Springer-Verlag) ; v. 129.
QA403.3.W363 1998
515'.2433--dc21   97-48855

Printed on acid-free paper.

Camera ready copy provided by the authors.
Printed and bound by Braun-Brumfield, Ann Arbor, MI.
Printed in the United States of America.

9 8 7 6 5 4 3 2

ISBN 0-387-98453-4 Springer-Verlag New York Berlin Heidelberg   SPIN 10754287

Preface

The mathematical theory of *ondelettes* (wavelets) was developed by Yves Meyer and many collaborators about 10 years ago. It was designed for approximation of possibly irregular functions and surfaces and was successfully applied in data compression, turbulence analysis, image and signal processing. Five years ago wavelet theory progressively appeared to be a powerful framework for nonparametric statistical problems. Efficient computational implementations are beginning to surface in this second lustrum of the nineties. This book brings together these three main streams of wavelet theory. It presents the theory, discusses approximations and gives a variety of statistical applications. It is the aim of this text to introduce the novice in this field into the various aspects of wavelets. Wavelets require a highly interactive computing interface. We present therefore all applications with software code from an interactive statistical computing environment.

Readers interested in theory and construction of wavelets will find here in a condensed form results that are somewhat scattered around in the research literature. A practioner will be able to use wavelets via the available software code. We hope therefore to address both theory and practice with this book and thus help to construct bridges between the different groups of scientists.

This text grew out of a French-German cooperation (*Séminaire Paris-Berlin, Seminar Berlin-Paris*). This seminar brings together theoretical and applied statisticians from Berlin and Paris. This work originates in the first of these seminars organized in Garchy, Burgundy in 1994. We are confident that there will be future research work originating from this yearly seminar.

This text would not have been possible without discussion and encouragement from colleagues in France and Germany. We would like to thank in particular Lucien Birgé, Christian Gourieroux, Yuri Golubev, Marc Hoffmann, Sylvie Huet, Emmanuel Jolivet, Oleg Lepski, Enno Mammen, Pascal

Massart, Michael Nussbaum, Michael Neumann, Volodja Spokoiny, Karine Tribouley. The help of Yuri Golubev was particularly important. Our Sections 11.5 and 12.5 are inspired by the notes that he kindly provided. The implementation in XploRe was professionally arranged by Sigbert Klinke and Clementine Dalelane. Steve Marron has established a fine set of test functions that we used in the simulations. Michael Kohler and Marc Hoffmann made many useful remarks that helped in improving the presentation. We had strong help in designing and applying our LaTeX macros from Wolfram Kempe, Anja Bardeleben, Michaela Draganska, Andrea Tiersch and Kerstin Zanter. Un très grand merci!

Berlin-Paris, September 1997

Wolfgang Härdle
Gerard Kerkyacharian,
Dominique Picard
Alexander Tsybakov

Contents

List of Figures

List of Tables

Symbols and Notation

φ	father wavelet
ψ	mother wavelet
$S1, S2, \ldots$	symmlets
$D1, D2, \ldots$	Daubechies wavelets
$C1, C2, \ldots$	Coiflets
ISE	integrated squared error
$MISE$	mean integrated squared error
$I\!\!R$	the real line
$Z\!\!\!Z$	set of all integers in $I\!\!R$
l_p	space of p-summable sequences
$L_p(I\!\!R)$	space of p-integrable functions
$W_p^m(I\!\!R)$	Sobolev space
$B_p^{sq}(I\!\!R)$	Besov space
$\mathcal{D}(I\!\!R)$	space of infinitely many times differentiable compactly supported functions
$S'(I\!\!R)$	Schwartz space
H^λ	Hölder smoothness class with parameter λ
(f, g)	scalar product in $L_2(I\!\!R)$
$||f||_p$	norm in $L_p(I\!\!R)$
$||a||_{l_p}$	norm in l_p
$||f||_{spq}$	norm in $B_p^{sq}(I\!\!R)$
ONS	orthonormal system
ONB	orthonormal basis
MRA	multiresolution analysis
RHS	right hand side
LHS	left hand side
DWT	discrete wavelet transform

$f * g$	convolution of f and g
$I\{A\}$	indicator function of a set A
a.e.	almost everywhere
$supp\ f$	support of function f
ess sup	essential supremum
$f^{(m)}$	m-th derivative
$\tau_h f(x) = f(x - h)$	shift operator
$\omega_p^1(f, t)$	modulus of continuity in the L_p norm
$K(x, y)$	kernel
δ_{jk}	Kronecker's delta
$\simeq$	asymptotic identical rate
$\sum_k$	sum over all $k \in \mathbb{Z}$
card Ω	cardinality of a set Ω

Chapter 1

Wavelets

1.1 What can wavelets offer?

A wavelet is, as the name suggests, a small wave. Many statistical phenomena have wavelet structure. Often small bursts of high frequency wavelets are followed by lower frequency waves or vice versa. The theory of wavelet reconstruction helps to localize and identify such accumulations of small waves and helps thus to better understand reasons for these phenomena. Wavelet theory is different from Fourier analysis and spectral theory since it is based on a local frequency representation.

Let us start with some illustrative examples of wavelet analysis for financial time series data. Figure 1.1 shows the time series of 25434 log(ask) – log(bid) spreads of the DeutschMark (DEM) - USDollar (USD) exchange rates during the time period of October 1, 1992 to September 30, 1993. The series consists of offers (bids) and demands (asks) that appeared on the FXFX page of the Reuters network over the entire year, see Bossaerts, Hafner & Härdle (1996), Ghysels, Gourieroux & Jasiak (1995). The graph shows the bid - ask spreads for each quarter of the year on the vertical axis. The horizontal axis denotes time for each quarter.

The quarterly time series show local bursts of different size and frequency. Figure 1.2 is a zoom of the first quarter. One sees that the bid-ask spread varies dominantly between 2 - 3 levels, has asymmetric behavior with thin but high rare peaks to the top and more oscillations downwards. Wavelets provide a way to quantify this phenomenon and thereby help to detect mechanisms

1

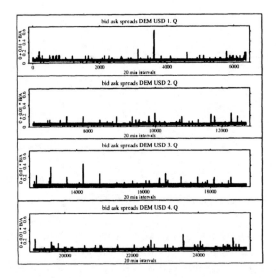

Figure 1.1: Bid-Ask spreads for one year of the DEM-USD FX-rate.

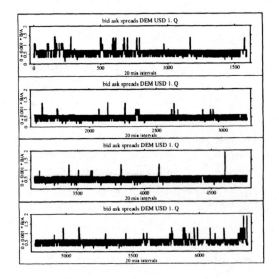

Figure 1.2: The first quarter of the DEM-USD FX rate.

for these local bursts.

Figure 1.3 shows the first 1024 points (about 2 weeks) of this series in the upper plot and the size of "wavelet coefficients" in the lower plot. The definition of wavelet coefficients will be given in Chapter 3. Here it suffices to view them as the values that quantify the location, both in time and frequency domain, of the important features of the function.

The lower half of Figure 1.3 is called *location - frequency plot* . It is interpreted as follows. The Y–axis contains four levels (denoted by 2,3,4 and 5) that correspond to different frequencies. Level 5 and level 2 represent the highest and the lowest frequencies respectively. The X–axis gives the location in time. The size of a bar is proportional to the absolute value of the wavelet coefficient at the corresponding level and time point. The lowest frequency level 2 chops this two week time interval into 4 half weeks. We recognize a high activity in the first half week. The next level 3 (8 time intervals) brings up a high activity peak after 2 days. The next higher level (roughly one day per interval) points us to two active days in this week.

In Figure 1.4 we represent in the same scale as Figure 1.3 the wavelet coefficients for the next 1024 points, again a two week interval. We see in comparison with the first two weeks that this time the activity is quite different: the bid-ask spread has smaller values that vary more regularly.

Let us compare this DEM/USD foreign exchange pattern with the exchange between Japanese YEN and DEM. Figure 1.5 shows the plot corresponding to Figure 1.3. We see immediately from the wavelet coefficients that the daily activity pattern is quite different on this market. An application of wavelet techniques to jump detection for monthly stock market return data is given in Wang (1995), see also Raimondo (1996).

A Fourier frequency spectrum would not be able to represent these effects since it is not sensitive to effects that are local in time. Figure 1.7 shows the estimated Fourier frequency spectral density for the YEN/DEM series of Figure 1.6. Note that the symmetric center of this graph corresponds to waves of a week's length. We see the high frequency of a one day activity as in the uppermost level of Figure 1.5, but not when this happens. Wavelets provide a spatial frequency resolution, whereas the Fourier frequency representation gives us only a global, space insensitive frequency distribution. (In our univariate example "space" corresponds to time.)

The spatial sensitivity of wavelets is useful also in smoothing problems, in particular in density and regression estimation. Figure 1.8 shows two

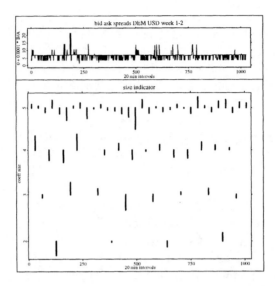

Figure 1.3: The first 1024 points (2 weeks) of the DEM-USD FX rate with a location - frequency plot.

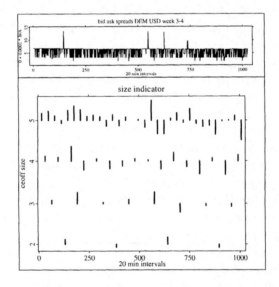

Figure 1.4: Distribution of coefficients for weeks 3–4.

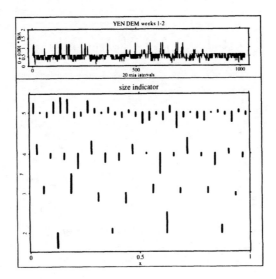

Figure 1.5: The first 2 weeks of the YENDEM FX-rate.

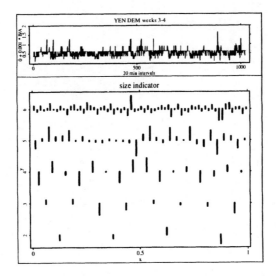

Figure 1.6: The weeks 3 - 4 of the YENDEM FX-rate.

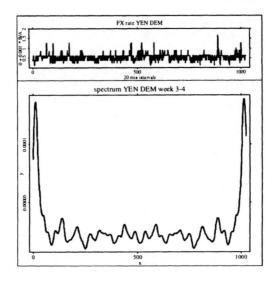

Figure 1.7: The smoothed periodogram of the YENDEM series.

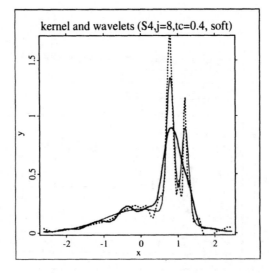

Figure 1.8: Binned Belgian household data at x–axis. Wavelet density estimate (solid) and kernel density estimate (dashed).

estimates of a total expenditure density for Belgian households. The dotted line is a kernel density estimate and the solid line a smoothed wavelet density estimate of the binned data given in the lower graph.

The kernel density estimate was computed with a Quartic kernel and the Silverman rule of thumb, see Silverman (1986), Härdle (1990). The binned data - a histogram with extremely small binwidth - shows a slight shoulder to the right corresponding to a possible mode in the income distribution. The kernel density estimate uses one single, global bandwidth for this data and is thus not sensitive to local curvature changes, like modes, troughs and sudden changes in the form of the density curve. One sees that the wavelet density estimate picks up two shoulders and models also the more sparsely distributed observations in the right tail of the distribution. This local smoothing feature of wavelets applies also to regression problems and will be studied in Chapter 10.

In summary, wavelets offer a frequency/time representation of data that allows us time (respectively, space) adaptive filtering, reconstruction and smoothing.

1.2 General remarks

The word "wavelet" is used in mathematics to denote a kind of orthonormal bases in L_2 with remarkable approximation properties. The theory of wavelets was developed by Y.Meyer, I.Daubechies, S.Mallat and others in the end of 1980-ies.

Qualitatively, the difference between the usual sine wave and a wavelet may be described by the localization property: the sine wave is localized in frequency domain, but not in time domain, while a wavelet is localized both in frequency and time domain. Figure 1.9 explains this difference. In the upper half of Figure 1.9 the sine waves $\sin(8\pi x)$, $\sin(16\pi x)$, $x \in (0,1)$ are shown. The frequency is stable over the horizontal axis, the "time" axis. The lower half of Figure 1.9 shows a typical example of two wavelets (Daubechies 10, denoted as $D10$, see Chapter 7). Here the frequency "changes" in horizontal direction.

By saying "localized" frequency we do not mean that the support of a wavelet is compact. We rather mean that the mass of oscillations of a wavelet is concentrated on a small interval. Clearly this is not the case for a sine wave.

The Fourier orthonormal basis is composed of waves, while the aim of the theory of wavelets is to construct orthonormal bases composed of wavelets.

Besides the already discussed localization property of wavelets there are other remarkable features of this technique. Wavelets provide a useful tool in data compression and have excellent statistical properties in data smoothing. This is shortly presented in the following sections.

1.3   Data compression

Wavelets allow to simplify the description of a complicated function in terms of a small number of coefficients. Often there are less coefficients necessary than in the classical Fourier analysis.

EXAMPLE 1.1 Define the step function

$$f(x) = \left\{ \begin{array}{ll} -1, & x \in \left[-\frac{1}{2}, 0\right], \\ 1, & x \in \left(0, \frac{1}{2}\right]. \end{array} \right.$$

This function is poorly approximated by its Fourier series. The Fourier expansion for $f(x)$ has the form

$$f(x) = \sum_{\substack{k=1 \\ k \text{ odd}}}^{\infty} \frac{4}{\pi k} \sin(2\pi k x) = \sum_{\substack{k=1 \\ k \text{ odd}}}^{\infty} c_k \varphi_k(x), \qquad (1.1)$$

where $\varphi_k(x) = \sqrt{2}\sin(2\pi k x)$ and $c_k = \frac{2\sqrt{2}}{\pi k}$. Figure 1.10 shows this function together with the approximated Fourier series with 5 terms. The Fourier coefficients c_k decrease as $O(k^{-1})$ which is a slow rate. So, one needs many terms of the Fourier expansion to approximate f with a good accuracy. Figure 1.11 shows the step function $f(x)$ with the Fourier expansion using 50 terms in (1.1). If we include 500 terms in this Fourier expansion it would not look drastically different from what we already see in Figure 1.11. The Fourier basis tends to keep the undesirable oscillations near the jump point and the endpoints of the interval.

Wavelets are more flexible. In fact, wavelet systems localize the jump by putting a small and extremely oscillating wavelet around the jump. This

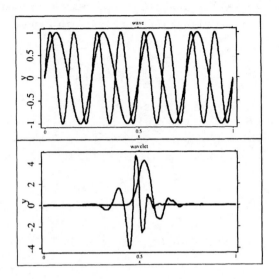

Figure 1.9: Sine and cosine waves and wavelets (D10).

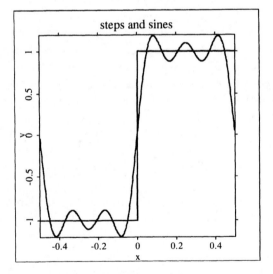

Figure 1.10: The step function and the Fourier series approximation with 5 terms.

involves only one (or small number) of coefficients, in contrast to the Fourier case. One such wavelet system is the Haar basis with (mother) wavelet

$$\psi(x) = \begin{cases} -1, & x \in [0, \frac{1}{2}], \\ 1, & x \in (\frac{1}{2}, 1]. \end{cases} \qquad (1.2)$$

The Haar basis consists of functions $\psi_{jk}(x) = 2^{j/2}\psi(2^j x - k)$, $j, k = \ldots,$ $-1, 0, 1, \ldots$. It is clear that with such a basis the step function in Figure 1.11 can be perfectly represented by two coefficients whereas using a Fourier series with 50 terms still produces wiggles in the reconstruction.

EXAMPLE 1.2 Let $f(x)$ be of the form shown in Figure 1.12. The function is

$$f(x) = I\{x \in [0, 0.5]\}\sin(8\pi x) + I\{x \in (0.5, 1]\}\sin(32\pi x)$$

sampled at $n = 512$ equidistant points. Here $I\{\cdot\}$ denotes the indicator function. That is, the support of f is composed of two intervals $[a, b] = [0, 0.5]$ and $[c, d] = [0.5, 1]$. On $[a, b]$ the frequency of oscillation of f is smaller than on $[c, d]$. If doing the Fourier expansion, one should include both frequencies: ω_1-,,frequency of $[a, b]$" and ω_2-,,frequency of $[c, d]$". But since the sine waves have infinite support, one is forced to compensate the influence of ω_1 on $[c, d]$ and of ω_2 on $[a, b]$ by adding a large number of higher frequency terms in the Fourier expansion. With wavelets one needs essentially only two pairs of time-frequency coefficients: $(\omega_1, [a, b])$ and $(\omega_2, [c, d])$. This is made clear in Figure 1.13 where we show a time frequency resolution as in Figure 1.3.

One clearly sees the dominant low frequency waves in the left part as high valued coefficients in Level 3 in the upper part of the graph. The highest frequency components occur in level 5. The sine wave was sampled at $n = 512$ points.

Figure 1.14 shows a wavelet approximation of the above sine wave example. The approximation is based on exactly the coefficients we see in the location - frequency plot in the lower part of Figure 1.14. Altogether only 18 coefficients are used to reconstruct the curve at $n = 512$ points. The reconstructed curve looks somewhat jagged due to the fact that we used a non smooth (so called $D4$) wavelet basis. We discuss later in Chapters 8 and 9 how to improve the approximation. The 18 coefficients were selected

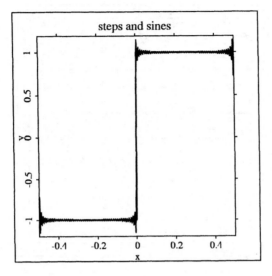

Figure 1.11: The step function and the Fourier series with 50 terms.

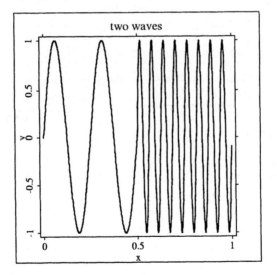

Figure 1.12: Two waves with different frequency.

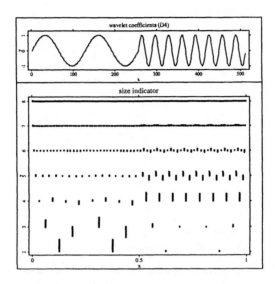

Figure 1.13: Location - frequency plot for the curve in Figure 1.12

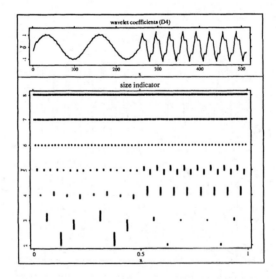

Figure 1.14: The wavelet approximation (with its location - frequency plot) for the curve of Figure 1.12

so that their absolute value was bigger than 0.4 times the maximal absolute coefficient value. We see that 18 coefficients suffice to reconstruct the curve at 512 points. This corresponds to a data compression rate of about $\frac{1}{32}$.

Wavelet data compression is especially useful in image processing, restoration and filtering. Consider an example. Figure 1.15 shows the Paris–Berlin seminar label on a grid of 256×256 points.

The picture was originally taken with a digital camera and discretized onto this grid. The original picture, as given on the front page of this text, has thus 65536 = 256 × 256 points. The image in Figure 1.15 was computed from only 500 coefficients (with Haar wavelets). This corresponds to a data compression rate of about 1/130. The shape of the picture is clearly visible, the text "séminaire Paris– Berlin" and "Seminar Berlin–Paris", though, is slightly disturbed but still readable at this level of compression.

1.4 Local adaptivity

This property was evident for the Examples 1.1 and 1.2. Wavelets are adapted to local properties of functions to to a larger extent than the Fourier basis. The adaptation is done automatically in view of the existence of a "second degree of freedom": the localization in time (or space, if multivariate functions are considered). We have seen in Figures 1.3, 1.4 and the above sine examples that wavelets represent functions and data both in levels (degree of resolution) and time. The vertical axis in these graphs denotes always the level, i.e. the partition of the time axis into finer and finer resolutions. In Figure 1.13 for example we saw that at level 3, corresponding to $2^3 = 8$ subintervals of the time interval [0,1], the low frequency part of the sine waves shows up. The higher frequencies appear only at level 5 when we divide [0,1] into $2^5 = 32$ subintervals. The advantage of this "multiresolution analysis" is that we can see immediately local properties of data and thereby influence our further analysis. The local form of the Belgian income distribution density for example becomes more evident when using wavelet smoothing, see Figure 1.8. Further examples are given in Chapters 10, 12.

There were attempts in the past to modify the Fourier analysis by partitioning the time domain into pieces and applying different Fourier expansions on different pieces. But the partitioning is always subjective. Wavelets pro-

Figure 1.15: The seminar label computed from 500 coefficients.

vide an elegant and mathematically consistent realization of this intuitive idea.

1.5 Nonlinear smoothing properties

The smoothing property of wavelets has been shortly mentioned above in the Belgian income estimation. In terms of series representations of functions smoothing means that we set some coefficients in this series equal to zero. This can be done in different ways. One way is to cut the series, starting from some prescribed term, for example, to keep only the first five terms of the expansion. This yields a traditional linear smoother (it is linear with respect to the coefficients of the series expansion). Another way is to keep only those coefficients, whose absolute value is greater than some threshold. The result is then a nonlinear function of the coefficients, and we obtain an example of a nonlinear smoother. Such a nonlinear way is called *thresholding*. We shall discuss this technique as we go along. It will be seen later (Chapter 10) that linear smoothers cannot achieve the minimax rate in the case of nonhomogeneous or unknown regularity of the estimated function. Wavelet thresholding provides a way to automatically adapt to the regularity of the function to be estimated and to achieve the minimax rate.

The wavelet thresholding procedure was proposed by D. Donoho and I. Johnstone in the beginning of 1990-ies. It is a very simple procedure, and it may seem almost to be a miracle that it provides an answer to this hard mathematical problem.

1.6 Synopsis

This book is designed to provide an introduction to the theory and practice of wavelets. We therefore start with the simplest wavelet basis, the Haar basis (Chapter 2). Then we give the basic idea of space/frequency multiresolution analysis (Chapter 3) and we recall some facts from Fourier analysis (Chapter 4) related to the fixed frequency resolution theory.

The basics of wavelet theory are presented in Chapter 5 followed by a chapter on the actual construction of wavelets. Chapter 7 is devoted to Daubechies' construction of compactly supported wavelets. Chapters 8 and 9 study the approximation properties of wavelet decomposition and give an introduction to Besov spaces which correspond to an appropriate functional framework. In Chapter 10 we introduce some statistical wavelet estimation procedures and study their properties. Chapter 11 is concerned with the adaptation issue in wavelet estimation. The final Chapter 12 discusses computational aspects and an interactive software interface. In the appendix we give coefficients used to generate wavelets and the address for the XploRe software sources (Härdle, Klinke & Turlach (1995)).

Chapter 2

The Haar basis wavelet system

The Haar basis is known since 1910. Here we consider the Haar basis on the real line $I\!\!R$ and describe some of its properties which are useful for the construction of general wavelet systems. Let $L_2(I\!\!R)$ be the space of all complex valued functions f on $I\!\!R$ such that their L_2-norm is finite:

$$||f||_2 = \left(\int_{-\infty}^{\infty} |f(x)|^2 dx \right)^{1/2} < \infty.$$

This space is endowed with the scalar product

$$(f,g) = \int_{-\infty}^{\infty} f(x)\overline{g(x)}dx.$$

Here and later $\overline{g(x)}$ denotes the complex conjugate of $g(x)$. We say that $f, g \in L_2(I\!\!R)$ are orthogonal to each other if $(f,g) = 0$ (in this case we write $f \perp g$).

Note that in this chapter we deal with the space $L_2(I\!\!R)$ of complex-valued functions. This is done to make the argument consistent with the more general framework considered later. However, for the particular case of this chapter the reader may also think of $L_2(I\!\!R)$ as the space of real-valued functions, with no changes in the notation.

A system of functions $\{\varphi_k, \ k \in Z\!\!\!Z\}$, $\varphi_k \in L_2(I\!\!R)$, is called *orthonormal system* (ONS) if

$$\int \varphi_k(x)\overline{\varphi_j(x)}dx = \delta_{jk},$$

17

where δ_{jk} is the Kronecker delta. An ONS $\{\varphi_k,\ k \in \mathbb{Z}\}$ is called *orthonormal basis* (ONB) in a subspace V of $L_2(\mathbb{R})$ if any function $f \in V$ has a representation

$$f(x) = \sum_k c_k \varphi_k(x),$$

where the coefficients c_k satisfy $\sum_k |c_k|^2 < \infty$. Here and later

$$\mathbb{Z} = \{\dots, -1, 0, 1, \dots\}, \quad \sum_k = \sum_{k=-\infty}^{\infty}, \quad \int = \int_{-\infty}^{\infty}.$$

Consider the following subspace V_0 of $L_2(\mathbb{R})$:

$$V_0 = \{f \in L_2(\mathbb{R}) : f \quad \text{is constant on} \quad (k, k+1], \quad k \in \mathbb{Z}\}.$$

Clearly,

$$f \in V_0 \iff f(x) = \sum_k c_k \varphi(x - k),$$

where $\sum_k |c_k|^2 < \infty$, the series converges in $L_2(\mathbb{R})$, and

$$\varphi(x) = I\{x \in (0,1]\} = \begin{cases} 1, & x \in (0,1], \\ 0, & x \notin (0,1]. \end{cases} \tag{2.1}$$

Denote

$$\varphi_{0k}(x) = \varphi(x - k), \qquad k \in \mathbb{Z}.$$

REMARK 2.1 The system $\{\varphi_{0k}\}$ is an orthonormal basis (ONB) in V_0.

Now, define a new linear subspace of $L_2(\mathbb{R})$ by

$$V_1 = \{h(x) = f(2x) : f \in V_0\}.$$

The space V_1 contains all functions in $L_2(\mathbb{R})$ that are constant on the intervals of the form $(\frac{k}{2}, \frac{k+1}{2}], \quad k \in \mathbb{Z}$.

Obviously, $V_0 \subset V_1$, and an ONB in V_1 is given by the system of functions $\{\varphi_{1k}\}$, where

$$\varphi_{1k}(x) = \sqrt{2}\varphi(2x - k), \qquad k \in \mathbb{Z}.$$

One can iterate this process and define, in general, the space

$$V_j = \{h(x) = f(2^j x) : f \in V_0\}.$$

Then V_j is a linear subspace of $L_2(I\!R)$ with the ONB

$$\varphi_{jk}(x) = 2^{j/2}\varphi(2^j x - k), \qquad k \in Z,$$

and

$$V_0 \subset V_1 \subset \ldots \subset V_j \subset \ldots$$

In the same way one defines the spaces V_j for $j < 0, j \in Z$, and one gets the inclusions

$$\ldots \subset V_{-1} \subset V_0 \subset V_1 \subset \ldots$$

Continuing this process infinitely, we approximate the whole space $L_2(I\!R)$.

PROPOSITION 2.1 $\bigcup_{j=0}^{\infty} V_j$ *(and hence* $\bigcup_{j=-\infty}^{\infty} V_j$ *) is dense in* $L_2(I\!R)$.

Proof follows immediately from the fact that every $f \in L_2(I\!R)$ can be approximated by a piecewise constant function $\tilde{f} \in L_2(I\!R)$ of the form $\sum_m \tilde{c}_m I\{x \in A_m\}$ where A_m are intervals, and each $I\{x \in A_m\}$ may be approximated by a sum of indicator functions of intervals of the form $(\frac{k}{2^j}, \frac{k+1}{2^j}]$. In other words, linear span of the system of functions $\{\{\varphi_{0k}\}, \{\varphi_{1k}\}, \ldots\}$ is dense in $L_2(I\!R)$. Clearly, this system is not a basis in $L_2(I\!R)$. But it can be transformed to a basis by means of orthogonalization. How to orthogonalize it?

Denote by W_0 the orthogonal complement of V_0 in V_1:

$$W_0 = V_1 \ominus V_0.$$

(In other terms, $V_1 = V_0 \oplus W_0$). This writing means that every $v_1 \in V_1$ can be represented as $v_1 = v_0 + w_0, \quad v_0 \in V_0, \quad w_0 \in W_0$, where $v_0 \perp w_0$. How to describe the space W_0 ? Let us show that W_0 is a linear subspace of $L_2(I\!R)$ spanned by a certain ONB. This will answer the question. Pick the following function

$$\psi(x) = \begin{cases} -1, & x \in [0, \frac{1}{2}], \\ 1, & x \in (\frac{1}{2}, 1]. \end{cases} \tag{2.2}$$

PROPOSITION 2.2 *The system* $\{\psi_{0k}\}$ *where*

$$\psi_{0k}(x) = \psi(x - k), \qquad k \in Z,$$

is an ONB in W_0. In other terms, W_0 is the linear subspace of $L_2(\mathbb{R})$ which is composed of the functions of the form

$$f(x) = \sum_k c_k \psi(x - k)$$

where $\sum_k |c_k|^2 < \infty$, and the series converges in $L_2(\mathbb{R})$.

Proof It suffices to verify the following 3 facts:

(i) *$\{\psi_{0k}\}$ is an orthonormal system (ONS).* This is obvious, since the supports of ψ_{0l} and ψ_{0k} are non-overlapping for $l \neq k$, and $||\psi_{0k}||_2 = 1$.

(ii) *$\{\psi_{0k}\}$ is orthogonal to V_0*, i.e.

$$(\psi_{0k}, \varphi_{0l}) = \int \psi_{0k}(x)\varphi_{0l}(x)dx = 0, \qquad \forall l, k.$$

If $l \neq k$, this is trivial (non-overlapping supports of ψ_{0k} and φ_{0l}). If $l = k$, this follows from the definition of ψ_{0k}, φ_{0k}:

$$\int \psi_{0k}(x)\varphi_{0k}(x)dx = \int_0^1 \psi(x)\varphi(x)dx = \int_0^1 \psi(x)dx = 0.$$

(iii) *Every $f \in V_1$ has a unique representation in terms of the joint system $\{\{\varphi_{0k}\}, \{\psi_{0k}\}, k \in \mathbb{Z}\}$.*

Let $f \in V_1$. Then

$$f(x) = \sum_k c_k \varphi_{1k}(x), \qquad \sum_k |c_k|^2 < \infty.$$

This representation is unique since $\{\varphi_{1k}\}$ is an ONB in V_1. Thus, it suffices to prove that φ_{1k} is a linear combination of φ_{0k} and ψ_{0k} for each k. It suffices to consider the case where $k = 0$ and $k = 1$. One easily shows that

$$
\begin{aligned}
\varphi_{10}(x) &= \sqrt{2}\varphi(2x) = \sqrt{2}\, I\{x \in (0, \tfrac{1}{2}]\} \\
&= \sqrt{2}\{\varphi_{00}(x) - \psi_{00}(x)\}/2 = \frac{1}{\sqrt{2}}\{\varphi_{00}(x) - \psi_{00}(x)\}
\end{aligned}
$$

Similarly, $\varphi_{11}(x) = \sqrt{2}\varphi(2x-1) = \frac{1}{\sqrt{2}}\{\varphi_{00}(x)+\psi_{00}(x)\}$.

We have $V_1 = V_0 \oplus W_0$. One can extend this construction to every V_j, to get

$$V_{j+1} = V_j \oplus W_j$$

where $W_j = V_{j+1} \ominus V_j$ is the orthogonal complement of V_j in V_{j+1}. In particular, the system $\{\psi_{jk}, \ k \in \mathbb{Z}\}$, where $\psi_{jk}(x) = 2^{j/2}\psi(2^j x - k)$, is ONB in W_j. Formally, we can write this as:

$$V_{j+1} = V_j \oplus W_j = V_{j-1} \oplus W_{j-1} \oplus W_j = \ldots = V_0 \oplus W_0 \oplus W_1 \oplus \ldots \oplus W_j = V_0 \oplus \bigoplus_{l=0}^{j} W_l.$$

We know that $\bigcup_j V_j$ is dense in $L_2(\mathbb{R})$, or, in other terms,

$$\overline{\bigcup_j V_j} = L_2(\mathbb{R}).$$

Using the orthogonal sum decomposition of V_j, one gets also

$$L_2(\mathbb{R}) = V_0 \oplus \bigoplus_{j=0}^{\infty} W_j.$$

This symbolic writing means that every $f \in L_2(\mathbb{R})$ can be represented as a series (convergent in $L_2(\mathbb{R})$) of the form

$$f(x) = \sum_k \alpha_{0k}\varphi_{0k}(x) + \sum_{j=0}^{\infty}\sum_k \beta_{jk}\psi_{jk}(x) \tag{2.3}$$

where α_{0k}, β_{jk} are the coefficients of this expansion. For sake of simplicity we shall often use the notation α_k instead of α_{0k}.

COROLLARY 2.1 *The system of functions*

$$\left\{ \{\varphi_{0k}\}, \{\psi_{jk}\}, \ k \in \mathbb{Z}, \ j = 0, 1, 2, \ldots \right\}$$

is an ONB in $L_2(\mathbb{R})$.

REMARK 2.2 This representation is the one we used in the graphical displays of Chapter 1. The coefficients we showed in the upper part of the graphs were the coefficients β_{jk}.

REMARK 2.3 The expansion (2.3) has the property of localization both in time and frequency. In fact, the summation in k corresponds to localization in time (shifts of functions $\varphi_{j0}(x)$ and $\psi_{j0}(x)$). On the other hand, summation in j corresponds to localization in frequency domain. The larger is j, the higher is the "frequency" related to ψ_{jk}.

In fact, (2.3) presents a special example of *wavelet expansion*, which corresponds to our special choice of φ and ψ, given by (2.1) and (2.2). One may suppose that there exist other choices of φ and ψ which provide such expansion. This will be discussed later. The function φ is called *father wavelet*, ψ is *mother wavelet* (φ_{0k}, ψ_{jk} are "children").

REMARK 2.4 The mother wavelet ψ may be defined in a different way, for example

$$\psi(x) = \left\{ \begin{array}{rl} 1, & x \in [0, \frac{1}{2}], \\ -1, & x \in (\frac{1}{2}, 1]. \end{array} \right.$$

There are many functions which are orthogonal to φ, and one can choose ψ among these functions. (In fact, for a given father φ there may be several mothers ψ).

The situation of formula (2.3) is shown in Figure 2.1. We come back there to our sine wave Example 1.2 and approximate it by only a few terms of the Haar wavelet expansion.

More precisely, we use levels $j = 2, 3, 4$, and 18 non-zero coefficients β_{jk} shown in size in the lower part of the figure. The corresponding approximation is shown in the upper part of Figure 2.1. The high frequency part is nicely picked up but due to the simple step function form of this wavelet basis the smooth character of the sine wave is not captured. It is therefore interesting to look for other wavelet basis systems.

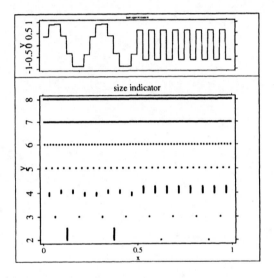

Figure 2.1: The sine example with a coarse Haar approximation.

Chapter 3

The idea of multiresolution analysis

3.1 Multiresolution analysis

The Haar system is not very convenient for approximation of smooth functions. In fact, any Haar approximation is a discontinuous function. One can show that even if the function f is very smooth, the Haar coefficients still decrease slowly. We therefore aim to construct wavelets that have better approximation properties.

Let φ be some function from $L_2(\mathbb{R})$, such that the family of translates of φ, i.e. $\{\varphi_{0k}, \ k \in \mathbb{Z}\} = \{\varphi(\cdot - k), \ k \in \mathbb{Z}\}$ is an *orthonormal system* (ONS). Here and later

$$\varphi_{jk}(x) = 2^{j/2}\varphi(2^j x - k), \ j \in \mathbb{Z}, \ k \in \mathbb{Z}.$$

Define the linear spaces

$$
\begin{aligned}
V_0 &= \{f(x) = \sum_k c_k \varphi(x - k) \ : \ \sum_k |c_k|^2 < \infty\}, \\
V_1 &= \{h(x) = f(2x) \ : \ f \in V_0\}, \\
&\vdots \\
V_j &= \{h(x) = f(2^j x) \ : \ f \in V_0\}, \ j \in \mathbb{Z}.
\end{aligned}
$$

We say that φ *generates the sequence of spaces* $\{V_j, j \in \mathbb{Z}\}$. Assume that

25

the function φ is chosen in such a way that the spaces are nested:

$$V_j \subset V_{j+1}, \qquad j \in \mathbb{Z}, \tag{3.1}$$

and that

$$\bigcup_{j \geq 0} V_j \text{ is dense in } \quad L_2(\mathbb{R}). \tag{3.2}$$

We proved in Chapter 2 that the relations (3.1) and (3.2) are satisfied for the Haar basis.

DEFINITION 3.1 *Let $\{\varphi_{0k}\}$ be an orthonormal system in $L_2(\mathbb{R})$. The sequence of spaces $\{V_j, j \in \mathbb{Z}\}$, generated by φ is called a* **multiresolution analysis** *(MRA) of $L_2(\mathbb{R})$ if it satisfies (3.1) and (3.2).*

The notion of multiresolution analysis was introduced by Mallat and Meyer in the years 1988–89 (see the books by Meyer(1990, 1993) and the article by Mallat (1989)). A link between multiresolution analysis and approximation of functions will be discussed in detail in Chapters 8 and 9.

DEFINITION 3.2 *If $\{V_j, \ j \in \mathbb{Z}\}$, is a MRA of $L_2(\mathbb{R})$, we say that the function φ generates a MRA of $L_2(\mathbb{R})$, and we call φ the* **father wavelet***.*

Assume that $\{V_j, \ j \in \mathbb{Z}\}$ is a MRA. Define

$$W_j = V_{j+1} \ominus V_j, \qquad j \in \mathbb{Z}.$$

Then, as in the case of Haar basis, we get

$$V_j = V_0 \oplus \bigoplus_{l=0}^{j} W_l,$$

since (3.1) holds. Iterating this infinitely many times, we find

$$\bigcup_{j=0}^{\infty} V_j = V_0 \oplus \bigoplus_{j=0}^{\infty} W_j. \tag{3.3}$$

By (3.2) and (3.3) one obtains

$$L_2(\mathbb{R}) = V_0 \oplus \bigoplus_{j=0}^{\infty} W_j.$$

This means that any $f \in L_2(I\!R)$ can be represented as a series (convergent in $L_2(I\!R)$):

$$f(x) = \sum_k \alpha_k \varphi_{0k}(x) + \sum_{j=0}^{\infty} \sum_k \beta_{jk} \psi_{jk}(x), \qquad (3.4)$$

where α_k, β_{jk} are some coefficients, and $\{\psi_{jk}\}, k \in Z\!\!\!Z$, is a *basis* for W_j. Note that there is a difference between (2.3) and (3.4):

in (2.3) $\psi_{jk}(x) = 2^{j/2}\psi(2^j x - k)$, where ψ is defined by (2.2),

in (3.4) $\{\psi_{jk}(x)\}$ is a general basis for W_j.

The relation (3.4) is called a *multiresolution expansion of f*. To turn (3.4) into the *wavelet expansion* one needs to justify the use of

$$\psi_{jk}(x) = 2^{j/2}\psi(2^j x - k)$$

in (3.4), i.e. the existence of such a function ψ called *mother wavelet*.

The space W_j is called *resolution level* of multiresolution analysis. In the Fourier analysis we have only one resolution level. In multiresolution analysis there are many resolution levels which is the origin of its name.

In the following, by abuse of notation, we frequently write "resolution level j" or simply "level j". We employ these words mostly to designate not the space W_j itself, but rather the coefficients β_{jk} and the functions ψ_{jk} "on the level j".

3.2 Wavelet system construction

The general framework of wavelet system construction looks like this:

1. Pick a function φ (*father wavelet*) such that $\{\varphi_{0k}\}$ is an orthonormal system, and (3.1), (3.2) are satisfied, i.e. φ generates a MRA of $L_2(I\!R)$.

2. Find a function $\psi \in W_0$ such that $\{\psi_{0k}, \ k \in Z\!\!\!Z\} = \{\psi(\cdot - k), \ k \in Z\!\!\!Z\}$, is ONB in W_0. This function is called *mother wavelet*. Then, consequently, $\{\psi_{jk}, \ k \in Z\!\!\!Z\}$ is ONB in W_j. Note that the mother wavelet is always orthogonal to the father wavelet.

3. Conclude that any $f \in L_2(I\!R)$ has the unique representation in terms of an L_2-convergent series:

$$f(x) = \sum_k \alpha_k \varphi_{0k}(x) + \sum_{j=0}^{\infty} \sum_k \beta_{jk} \psi_{jk}(x), \qquad (3.5)$$

where the *wavelet coefficients* are

$$\alpha_k = \int f(x)\overline{\varphi_{0k}(x)}dx, \qquad \beta_{jk} = \int f(x)\overline{\psi_{jk}(x)}dx.$$

The relation (3.5) is then called *inhomogeneous wavelet expansion*. One may also consider the *homogeneous wavelet expansion*

$$f(x) = \sum_{j=-\infty}^{\infty} \sum_k \beta_{jk} \psi_{jk}(x),$$

where the "reference" space V_0 is eliminated. The α_k coefficients summarize the general form of the function and the β_{jk} represent the innovations to this general form, the local details. This is why the β_{jk} are often called *detail coefficients*.

The fact that the expansion (3.5) starts from the reference space V_0 is just conventional. One can also choose V_{j_0}, for some $j_0 \in Z\!\!\!Z$, in place of V_0. Then the inhomogeneous wavelet expansion is of the form

$$f(x) = \sum_k \alpha_{j_0 k} \varphi_{j_0 k}(x) + \sum_{j=j_0}^{\infty} \sum_k \beta_{jk} \psi_{jk}(x),$$

where

$$\alpha_{jk} = \int f(x)\overline{\varphi_{jk}(x)}dx.$$

In the following (up to Chapter 9) we put $j_0 = 0$ to simplify the notation.

An immediate consequence of the wavelet expansion is that the orthogonal projection $P_{V_{j+1}}(f)$ of f onto V_{j+1} is of the form

$$P_{V_{j+1}}(f) = \sum_k \alpha_{j+1,k} \varphi_{j+1,k}(x) = \sum_k \alpha_{jk} \varphi_{jk}(x) + \sum_k \beta_{jk} \psi_{jk}(x). \qquad (3.6)$$

3.3 An example

Besides the Haar wavelet example considered in Chapter 2, another classical example of multiresolution analysis can be constructed via the *Shannon basis*. In this case the space $V_0 = V_0^{Sh}$ consists of functions $f \in L_2(I\!R)$ such that the Fourier transforms $\hat{f}(\xi)$ have support in $[-\pi, \pi]$. The space V_0^{Sh} is very famous in signal processing because of the following result (see for instance Papoulis (1977)).

Sampling theorem. *A function f belongs to V_0^{Sh} if and only if*

$$f(x) = \sum_k f(k) \frac{\sin \pi(x-k)}{\pi(x-k)}.$$

In words, the function $f \in V_0^{Sh}$ can be entirely recovered from its sampled values $\{f(k), \ k \in Z\!\!\!Z\}$.

It follows from the sampling theorem that the space $V_0 = V_0^{Sh}$ is generated by the function

$$\varphi(x) = \frac{\sin \pi x}{\pi x}. \tag{3.7}$$

The Fourier transform of φ is $\hat{\varphi}(\xi) = I\{\xi \in [-\pi, \pi]\}$. It is easy to see that the integer translates of φ form an ONS and that φ generates a MRA of $L_2(I\!R)$. In other words, φ defined in (3.7) is a father wavelet. The space V_j associated to this φ is the space of all functions in $L_2(I\!R)$ with Fourier transforms supported in $[-2^j\pi, 2^j\pi]$. This V_j is a space of very regular functions. It will be seen in Chapters 8 and 9 that projecting on V_j can be interpreted as a smoothing procedure.

We can also remark that in this example the coefficient of expansion has a special form since it is just the value $f(k)$. This situation is very uncommon, but some particular wavelets are constructed in such a way that the wavelet coefficients are "almost" interpolations of the function (e.g. coiflets, defined in Section 7.2).

Chapter 4

Some facts from Fourier analysis

This small chapter is here to summarize the classical facts of Fourier analysis that will be used in the sequel. We omit the proofs (except for the Poisson summation formula). They can be found in standard textbooks on the subject, for instance in Katznelson (1976), Stein & Weiss (1971).

Assume that $f \in L_1(I\!R)$, where $L_1(I\!R)$ is the space of all complex-valued functions f on $I\!R$, such that $\int_{-\infty}^{\infty} |f(x)| dx < \infty$. The *Fourier transform* of f is

$$\mathcal{F}[f](\xi) = \hat{f}(\xi) = \int_{-\infty}^{\infty} e^{-ix\xi} f(x) dx. \tag{4.1}$$

The function $\hat{f}$ is continuous and tends to zero when $|\xi| \to \infty$ (*Riemann-Lebesgue Lemma*). If $\hat{f}(\xi)$ is also absolutely integrable, there exists a continuous version of f and one can define the *inverse Fourier transform*

$$\mathcal{F}^{-1}[\hat{f}](x) = \frac{1}{2\pi} \int_{-\infty}^{\infty} e^{i\xi x} \hat{f}(\xi) d\xi, \tag{4.2}$$

and

$$f(x) = \frac{1}{2\pi} \int_{-\infty}^{\infty} e^{i\xi x} \hat{f}(\xi) d\xi = \mathcal{F}^{-1}[\hat{f}](x)$$

at almost every point x. In the following we assure that f is identical to its continuous version, whenever $\hat{f}(\xi)$ is absolutely integrable. Thus, in particular, the last equality holds for every x.

Recall the following properties of Fourier transform which are well known.

31

Plancherel formulas. If $f \in L_1(\mathbb{R}) \cap L_2(\mathbb{R})$, then

$$||f||_2^2 = \frac{1}{2\pi} \int_{-\infty}^{\infty} |\hat{f}(\xi)|^2 d\xi, \tag{4.3}$$

$$(f, g) = \frac{1}{2\pi} \int_{-\infty}^{\infty} \hat{f}(\xi)\overline{\hat{g}(\xi)} d\xi. \tag{4.4}$$

By extension, the Fourier transform can be defined for any $f \in L_2(\mathbb{R})$. In fact, the space $L_1(\mathbb{R}) \cap L_2(\mathbb{R})$ is dense in $L_2(\mathbb{R})$. Hence, by isometry (up to the factor $\frac{1}{2\pi}$) we define $\mathcal{F}[f]$ for any $f \in L_2(\mathbb{R})$, and (4.3) and (4.4) remain true for any $f, g \in L_2(\mathbb{R})$.

Fourier transform of a shifted function and scaled function.

$$\mathcal{F}[f(x - k)](\xi) = \int e^{-ix\xi} f(x - k)dx = e^{-ik\xi}\hat{f}(\xi). \tag{4.5}$$

$$\forall\, a > 0: \quad \mathcal{F}[f(ax)](\xi) = \int e^{-ix\xi} f(ax)dx = \frac{1}{a}\hat{f}\left(\frac{\xi}{a}\right). \tag{4.6}$$

Convolution.

We write $h = f * g$ for the convolution

$$h(x) = \int f(x - t)g(t)dt, \tag{4.7}$$

defined for any pair of functions f and g such that the RHS of this formula exists a.e. It is well known that in the frequency domain we have $\hat{h}(\xi) = \hat{f}(\xi)\hat{g}(\xi)$, if all the Fourier transforms in this formula exist.

Let $\tilde{f}(x) = \bar{f}(-x)$. Then

$$\mathcal{F}[f * \tilde{f}](\xi) = |\hat{f}(\xi)|^2. \tag{4.8}$$

Derivation.

If f is such that $\int |x|^N |f(x)|dx < \infty$, for some integer $N \geq 1$, then

$$\frac{d^N}{d\xi^N}\hat{f}(\xi) = \int f(t)(-it)^N \exp(-i\xi t)dt. \tag{4.9}$$

Conversely, if $\int |\xi|^N |\hat{f}(\xi)|d\xi < \infty$, then

$$(i\xi)^N \hat{f}(\xi) = \mathcal{F}[f^{(N)}](\xi). \tag{4.10}$$

Moreover, the following lemma holds.

LEMMA 4.1 *If $\hat{f}^{(j)}(\xi)$ are absolutely integrable for $j = 0, \ldots, N$, then*

$$|x|^N |f(x)| \to 0, \text{ as } |x| \to \infty.$$

Fourier series.

Let f be a 2π-periodic function on $\mathbb{R}$. We shall write for brevity $f \in L_p(0, 2\pi)$ if

$$f(x)I\{x \in [0, 2\pi]\} \in L_p(0, 2\pi), \ p \geq 1.$$

Any 2π-periodic function f on $\mathbb{R}$, such that $f \in L_2(0, 2\pi)$, can be represented by its Fourier series convergent in $L_2(0, 2\pi)$:

$$f(x) = \sum_k c_k e^{ikx},$$

where the Fourier coefficients are given by

$$c_k = \frac{1}{2\pi} \int_0^{2\pi} f(x) e^{-ikx} dx.$$

Also, by periodicity, this holds for all $x \in \mathbb{R}$.

The **Poisson summation formula** is given in the following theorem.

THEOREM 4.1 *Let $f \in L_1(\mathbb{R})$. Then the series*

$$S(x) = \sum_l f(x + 2l\pi) \tag{4.11}$$

converges a.e. and belongs to $L_1(0, 2\pi)$. Moreover the Fourier coefficients of $S(x)$ are given by

$$c_k = \frac{1}{2\pi} \hat{f}(k) = \mathcal{F}^{-1}[f](-k). \tag{4.12}$$

Proof For the first part it is enough to prove that

$$\int_0^{2\pi} \sum_l |f(x + 2l\pi)| dx < \infty.$$

This follows from the equality of this term to $\int_{-\infty}^{\infty} |f(x)| dx$. For the second part we have to compute the Fourier coefficients

$$\frac{1}{2\pi} \int_0^{2\pi} \{\sum_l f(x + 2l\pi)\} e^{-ikx} dx.$$

By exchanging summation and integration we arrive at

$$\sum_l \frac{1}{2\pi} \int_0^{2\pi} f(x + 2l\pi)e^{-ikx} dx$$

$$= \sum_l \frac{1}{2\pi} \int_{2\pi l}^{2\pi(l+1)} f(u)e^{-iku} du$$

$$= \frac{1}{2\pi} \hat{f}(k).$$

$\square$

REMARK 4.1 A necessary and sufficient condition for S in (4.11) to be equal to 1 a.e. is $\mathcal{F}^{-1}[f](0) = 1$ and $\mathcal{F}^{-1}[f](k) = 0$, $k \in \mathbb{Z}\backslash\{0\}$.

More generally, if $f \in L_1(\mathbb{R})$ and $T > 0$, then $\sum_l f(x + lT)$ is almost everywhere convergent and defines a T-periodic function whose Fourier coefficients are given by

$$\frac{1}{T} \int_0^T \sum_l f(x + lT) \exp\left(-ixk\frac{2\pi}{T}\right) dx = \frac{1}{T}\hat{f}\left(\frac{2\pi}{T}k\right). \qquad (4.13)$$

Chapter 5

Basic relations of wavelet theory

5.1   When do we have a wavelet expansion?

Let us formulate in the exact form the conditions on the functions φ and ψ which guarantee that the wavelet expansion (3.5) holds. This formulation is connected with the following questions.

Question 5.1 *How can we check that $\{\varphi_{0k}\}$ is an ONS?*

Question 5.2 *What are the sufficient conditions for (3.1) (nestedness of V_j) to hold?*

Question 5.3 *What are the conditions for (3.2) to hold, i.e. when is $\bigcup_j V_j$ dense in $L_2(\mathbb{R})$?*

Question 5.4 *Can we find a function $\psi \in W_0$ such that $\{\psi_{0k},\ k \in \mathbb{Z}\}$ is an ONB in W_0?*

These questions will be answered in turn in this chapter. An answer to Question 5.1 is given by the following lemma.

LEMMA 5.1 *Let $\varphi \in L_2(\mathbb{R})$. The system of functions $\{\varphi_{0k},\ k \in \mathbb{Z}\}$ is an ONS if and only if*

$$\sum_k |\hat{\varphi}(\xi + 2\pi k)|^2 = 1 \qquad (a.e.). \tag{5.1}$$

Proof Denote $q = \varphi * \check{\varphi}$ where $\check{\varphi}(x) = \overline{\varphi(-x)}$. Then, by (4.8),

$$\sum_k |\hat{\varphi}(\xi + 2\pi k)|^2 = \sum_k \hat{q}(\xi + 2\pi k).$$

As $\hat{q} = |\hat{\varphi}|^2 \in L_1(I\!R)$, Theorem 4.1 shows that this series converges a.e., and its Fourier coefficients are $c_k = \mathcal{F}^{-1}[\hat{q}](-k) = q(-k)$. The orthonormality condition reads as

$$\int \varphi(x - k)\overline{\varphi(x - l)}dx = \delta_{kl}, \quad \text{where } \delta_{kl} = \begin{cases} 1 & \text{if } k = l, \\ 0 & \text{if } k \neq l, \end{cases}$$

or, equivalently, $\int \varphi(x)\overline{\varphi(x - k)}dx = \delta_{0k}$. This gives

$$q(k) = \int \check{\varphi}(k - x)\varphi(x)dx = \int \varphi(x)\overline{\varphi(x - k)}dx = \delta_{0k}.$$

Using the Fourier expansion and Remark 4.1, we get

$$\sum_k \hat{q}(\xi + 2\pi k) = \sum_k c_k e^{ik\xi} = \sum_k q(k)e^{-ik\xi} = \sum_k \delta_{0k}e^{-ik\xi} = 1 \ (a.e.).\ \square$$

Let us now consider Question 5.2. We need to investigate the nestedness of the spaces V_j.

PROPOSITION 5.1 *The spaces V_j are nested,*

$$V_j \subset V_{j+1}, \quad j \in Z\!\!Z,$$

if and only if there exists a 2π-periodic function $m_0(\xi), m_0 \in L_2(0, 2\pi)$, such that

$$\hat{\varphi}(\xi) = m_0\left(\frac{\xi}{2}\right)\hat{\varphi}\left(\frac{\xi}{2}\right) \qquad (a.e.). \tag{5.2}$$

It suffices to prove this proposition for $j = 0$. First, prove that (5.2) is a necessary condition. Assume that $V_0 \subset V_1$. Hence, $\varphi \in V_1$. The system

$$\{\sqrt{2}\varphi(2x - k)\}$$

is a basis in V_1, by definition of V_1. Therefore, there exists a sequence $\{h_k\}$, such that

$$\varphi(x) = \sqrt{2}\sum_k h_k\varphi(2x - k), \tag{5.3}$$

$$h_k = \sqrt{2}\int \varphi(x)\overline{\varphi(2x - k)}dx, \quad \sum_k |h_k|^2 < \infty.$$

Take the Fourier transform of both sides of (5.3). Then, by (4.5), (4.6)

$$\hat{\varphi}(\xi) = \frac{1}{\sqrt{2}} \sum_k h_k e^{-i\xi k/2} \hat{\varphi}\left(\frac{\xi}{2}\right) = m_0\left(\frac{\xi}{2}\right) \hat{\varphi}\left(\frac{\xi}{2}\right) \qquad (a.e.)$$

where

$$m_0(\xi) = \frac{1}{\sqrt{2}} \sum_k h_k e^{-i\xi k}.$$

Note that $m_0(\xi)$ is a 2π-periodic function belonging to $L_2(0, 2\pi)$. Let us now turn to the proof of the converse. We begin with the following lemma.

LEMMA 5.2 *Let $\{\varphi_{0k}\}$ be an ONS. Every 2π-periodic function m_0 satisfying (5.2) such that $m_0 \in L_2(0, 2\pi)$, also satisfies*

$$|m_0(\xi)|^2 + |m_0(\xi + \pi)|^2 = 1 \qquad (a.e.).$$

Proof By (5.2)

$$|\hat{\varphi}(2\xi + 2\pi k)|^2 = |m_0(\xi + \pi k)|^2 |\hat{\varphi}(\xi + \pi k)|^2.$$

Summing up in k and using the fact that $\{\varphi_{0k}\}$ is an ONS and m_0 is 2π-periodic we get by Lemma 5.1 that a.e.

$$\begin{aligned}
1 &= \sum_{k=-\infty}^{\infty} |m_0(\xi + \pi k)|^2 |\hat{\varphi}(\xi + \pi k)|^2 \\
&= \sum_{l=-\infty}^{\infty} |m_0(\xi + 2\pi l)|^2 |\hat{\varphi}(\xi + 2\pi l)|^2 \\
&\quad + \sum_{l=-\infty}^{\infty} |m_0(\xi + 2\pi l + \pi)|^2 |\hat{\varphi}(\xi + 2\pi l + \pi)|^2 \\
&= \sum_{l=-\infty}^{\infty} |\hat{\varphi}(\xi + 2\pi l)|^2 |m_0(\xi)|^2 + \sum_{l=-\infty}^{\infty} |\hat{\varphi}(\xi + 2\pi l + \pi)|^2 |m_0(\xi + \pi)|^2 \\
&= |m_0(\xi)|^2 + |m_0(\xi + \pi)|^2. \qquad \qquad \Box
\end{aligned}$$

A consequence of this lemma is that such a function m_0 is bounded. Let us now finish the proof of Proposition 5.1. It is clear that if we denote by $\hat{V}_0$ (respectively $\hat{V}_1$) the set of Fourier transforms of the functions of V_0 (respectively V_1) we have:

$$\hat{V}_0 = \{m(\xi)\hat{\varphi}(\xi) : m(\xi) \ 2\pi\text{-periodic}, m \in L_2(0, 2\pi)\},$$

$$\hat{V}_1 = \{m(\xi/2)\hat{\varphi}(\xi/2) : m(\xi) \ 2\pi\text{-periodic}, m \in L_2(0, 2\pi)\}.$$

Condition (5.2) implies that every function in $\hat{V}_0$ has the form $m(\xi)m_0(\xi/2)$ $\hat{\varphi}(\xi/2)$ and belongs to $\hat{V}_1$. In fact $m(2\xi)m_0(\xi)$ is a 2π-periodic function belonging to $L_2(0, 2\pi)$ since $m \in L_2(0, 2\pi)$, and m_0 is bounded due to the previous lemma.

REMARK 5.1 It is always true that

$$\bigcap_j V_j = \{0\},$$

where 0 denotes the zero function (see Cohen & Ryan (1995), Theorem 1.1, p. 12).

The answer to Question 5.3. will be given in Chapter 8. It will be shown that if φ is a father wavelet, i.e. if (5.1) and (5.2) hold, then $\bigcup_j V_j$ is dense in $L_2(I\!R)$ whenever φ satisfies a mild integrability condition (see Corollary 8.1).

The answer to Question 5.4 is given in

LEMMA 5.3 *Let φ be a father wavelet which generates a MRA of $L_2(I\!R)$ and let $m_0(\xi)$ be a solution of (5.2). Then the inverse Fourier transform ψ of*

$$\hat{\psi}(\xi) = m_1\left(\frac{\xi}{2}\right)\hat{\varphi}\left(\frac{\xi}{2}\right), \qquad (5.4)$$

where $m_1(\xi) = \overline{m_0(\xi + \pi)}e^{-i\xi}$, is a mother wavelet.

REMARK 5.2 In other words, the lemma states that $\{\psi_{0k}\}$ is an ONB in W_0.

Proof We need to prove the following 3 facts.

(i) $\{\psi_{0k}\}$ *is an ONS*, i.e. by Lemma 5.1

$$\sum_k |\hat{\psi}(\xi + 2\pi k)|^2 = 1 \qquad \text{(a.e.)}.$$

Let us show this equality. With Lemma 5.2 and 2π-periodicity of m_0
we obtain

$$
\begin{aligned}
\sum_k |\hat{\psi}(\xi + 2\pi k)|^2 &= \sum_k \left| m_1\left(\frac{\xi}{2} + \pi k\right) \right|^2 \left| \hat{\varphi}\left(\frac{\xi}{2} + \pi k\right) \right|^2 \\
&= \sum_k \left| m_0\left(\frac{\xi}{2} + \pi + \pi k\right) \right|^2 \left| \hat{\varphi}\left(\frac{\xi}{2} + \pi k\right) \right|^2 \\
&= \sum_{l=-\infty}^{\infty} \left| m_0\left(\frac{\xi}{2} + \pi + 2\pi l + \pi\right) \right|^2 \left| \hat{\varphi}\left(\frac{\xi}{2} + 2\pi l + \pi\right) \right|^2 \\
&\quad + \sum_{l=-\infty}^{\infty} \left| m_0\left(\frac{\xi}{2} + \pi + 2\pi l\right) \right|^2 \left| \hat{\varphi}\left(\frac{\xi}{2} + 2\pi l\right) \right|^2 \\
&= \sum_{k=-\infty}^{\infty} \left| \hat{\varphi}\left(\frac{\xi}{2} + 2\pi k\right) \right|^2 = 1 \qquad \text{(a.e.).}
\end{aligned}
$$

(ii) $\{\psi_{0k}\}$ *is orthogonal to* $\{\varphi_{0k}\}$, *i.e.*

$$
\int \varphi(x - k)\overline{\psi(x - l)}dx = 0, \quad \forall k, l.
$$

It suffices to show that

$$
\int \varphi(x)\overline{\psi(x - k)}dx = 0, \quad \forall k,
$$

or, equivalently,

$$
g(k) = \varphi * \tilde{\psi}(k) = 0, \quad \forall k,
$$

where $g = \varphi * \tilde{\psi}$, $\tilde{\psi}(x) = \overline{\psi(-x)}$. The Fourier transform of g is

$$
\hat{g} = \hat{\varphi}\hat{\tilde{\psi}} = \hat{\varphi}\overline{\hat{\psi}}.
$$

Applying the Poisson summation formula (Theorem 4.1) to $f = \hat{g}$, we
get that the Fourier coefficients of the function $S(\xi) = \sum_k \hat{g}(\xi + 2\pi k)$
are $\mathcal{F}^{-1}[\hat{g}](-k) = g(-k)$, $k \in \mathbb{Z}$. Thus, the condition $g(k) = 0$, $\forall k$, is
equivalent to $S(\xi) = 0$ (a.e.), or

$$
\sum_k \hat{\varphi}(\xi + 2\pi k)\overline{\hat{\psi}(\xi + 2\pi k)} = 0 \qquad \text{(a.e.).} \tag{5.5}
$$

It remains to check (5.5). With our definition of $\hat{\psi}$, and using (5.2), we get

$$\sum_k \hat{\varphi}(\xi + 2\pi k)\overline{\hat{\psi}(\xi + 2\pi k)}$$

$$= \sum_k \hat{\varphi}\left(\frac{\xi}{2} + \pi k\right) m_0\left(\frac{\xi}{2} + \pi k\right) \overline{\hat{\varphi}\left(\frac{\xi}{2} + \pi k\right) m_1\left(\frac{\xi}{2} + \pi k\right)}$$

$$= \sum_k \left|\hat{\varphi}\left(\frac{\xi}{2} + \pi k\right)\right|^2 m_0\left(\frac{\xi}{2} + \pi k\right) \overline{m_1\left(\frac{\xi}{2} + \pi k\right)}$$

$$= m_0\left(\frac{\xi}{2}\right) \overline{m_1\left(\frac{\xi}{2}\right)} + m_0\left(\frac{\xi}{2} + \pi\right) \overline{m_1\left(\frac{\xi}{2} + \pi\right)}.$$

Thus (5.5) is equivalent to

$$m_0(\xi)\overline{m_1(\xi)} + m_0(\xi + \pi)\overline{m_1(\xi + \pi)} = 0 \qquad (a.e.). \qquad (5.6)$$

It remains to note that (5.6) is true, since

$$m_0(\xi)\overline{m_1(\xi)} \quad + \quad m_0(\xi + \pi)\overline{m_1(\xi + \pi)}$$
$$= \quad m_0(\xi)e^{i\xi}m_0(\xi + \pi) + m_0(\xi + \pi)e^{i\xi + i\pi}m_0(\xi) = 0.$$

(iii) *Any function f from V_1 has a unique representation*

$$f(x) = \sum_k c_k\varphi(x - k) + \sum_k c'_k\psi(x - k)$$

where c_k, c'_k are coefficients such that $\sum_k |c_k|^2 < \infty, \sum_k |c'_k|^2 < \infty$.

In fact, any $f \in V_1$ has a unique representation in terms of the ONB $\{\varphi_{1k}, k \in \mathbb{Z}\}$, where $\varphi_{1k}(x) = \sqrt{2}\varphi(2x - k)$. In the Fourier domain one can express this as in the proof of Proposition 5.1:

$$\hat{f}(\xi) = q\left(\frac{\xi}{2}\right) \hat{\varphi}\left(\frac{\xi}{2}\right) \qquad (a.e.) \qquad (5.7)$$

where

$$q(\xi) = \frac{1}{\sqrt{2}} \sum_k q_k e^{-i\xi k}.$$

Now, (5.2) and (5.4) entail

$$\overline{m_0\left(\frac{\xi}{2}\right)}\hat{\varphi}(\xi) = \left|m_0\left(\frac{\xi}{2}\right)\right|^2 \hat{\varphi}\left(\frac{\xi}{2}\right),$$

$$\overline{m_1\left(\frac{\xi}{2}\right)}\hat{\psi}(\xi) = \left|m_1\left(\frac{\xi}{2}\right)\right|^2 \hat{\varphi}\left(\frac{\xi}{2}\right).$$

By summing up these two equalities one gets

$$\hat{\varphi}\left(\frac{\xi}{2}\right)\left[\left|m_0\left(\frac{\xi}{2}\right)\right|^2 + \left|m_1\left(\frac{\xi}{2}\right)\right|^2\right]$$

$$= \overline{m_0\left(\frac{\xi}{2}\right)}\hat{\varphi}(\xi) + \overline{m_1\left(\frac{\xi}{2}\right)}\hat{\psi}(\xi) \qquad (a.e.). \qquad (5.8)$$

Note that

$$\left|m_1\left(\frac{\xi}{2}\right)\right|^2 = \left|m_0(\frac{\xi}{2} + \pi)\right|^2.$$

Using this and Lemma 5.2, we get from (5.8)

$$\hat{\varphi}\left(\frac{\xi}{2}\right) = \overline{m_0\left(\frac{\xi}{2}\right)}\hat{\varphi}(\xi) + \overline{m_1\left(\frac{\xi}{2}\right)}\hat{\psi}(\xi) \qquad (a.e.).$$

Substitute this into (5.7):

$$\hat{f}(\xi) = q\left(\frac{\xi}{2}\right)\overline{m_0\left(\frac{\xi}{2}\right)}\hat{\varphi}(\xi) + q\left(\frac{\xi}{2}\right)\overline{m_1\left(\frac{\xi}{2}\right)}\hat{\psi}(\xi) \qquad (a.e.).$$

By passing back to the time domain, we deduce that f has the unique representation in terms of $\{\varphi_{0k}\}$ and $\{\psi_{0k}\}$. □

REMARK 5.3 The statement of Lemma 5.3 is true if we choose m_1 in more general form

$$m_1(\xi) = \theta(\xi)\overline{m_0(\xi + \pi)}e^{-i\xi},$$

where $\theta(\xi)$ is an arbitrary π-periodic function such that $|\theta(\xi)| = 1$. One can easily check it as an exercise.

5.2   How to construct mothers from a father

Let us draw some conclusions from the answers to Questions 5.1 to 5.4.

Conclusion 1: As soon as we know the father wavelet $\varphi(x)$, and hence $\hat{\varphi}(\xi)$, we can immediately construct a mother wavelet ψ with the help of Lemmas 5.2 and 5.3. Indeed, from (5.2) we have $m_0(\xi) = \hat{\varphi}(2\xi)/\hat{\varphi}(\xi)$ and, from (5.4),

$$\hat{\psi}(\xi) = \overline{m_0\left(\frac{\xi}{2} + \pi\right)} e^{-i\xi/2} \hat{\varphi}\left(\frac{\xi}{2}\right). \tag{5.9}$$

The mother wavelet ψ is found by the inverse Fourier transform of $\hat{\psi}$.

Conclusion 2: It is still not clear how to find a father wavelet φ, but we proved some useful formulae that may help. These formulae are [1]

$$\sum_k |\hat{\varphi}(\xi + 2\pi k)|^2 = 1,$$

$$\hat{\varphi}(\xi) = m_0\left(\frac{\xi}{2}\right) \hat{\varphi}\left(\frac{\xi}{2}\right),$$

where

$$|m_0(\xi)|^2 + |m_0(\xi + \pi)|^2 = 1,$$

and $m_0(\xi)$ is 2π-periodic, $m_0 \in L_2(0, 2\pi)$.

It will be shown in Proposition 8.6 that for all reasonable examples of father wavelets we should have $|\hat{\varphi}(0)| = |\int \varphi(x)dx| = 1$, which yields immediately

$$m_0(0) = 1$$

(cf. (5.2)).

By adding this condition to the previous ones, we obtain the following set of relations:

$$\sum_k |\hat{\varphi}(\xi + 2\pi k)|^2 = 1, \tag{5.10}$$

$$\hat{\varphi}(\xi) = m_0\left(\frac{\xi}{2}\right) \hat{\varphi}\left(\frac{\xi}{2}\right), \tag{5.11}$$

---

[1]In the sequel we assume that $\hat{\varphi}$ and m_0 are continuous, so that we drop (a.e.) in all the relations.

and

$$|m_0(\xi)|^2 + |m_0(\xi + \pi)|^2 = 1, \\ m_0 \text{ is } 2\pi\text{-periodic}, \quad m_0 \in L_2(0, 2\pi), \\ m_0(0) = 1. \qquad\qquad (5.12)$$

The relations (5.9) – (5.12) provide a set of sufficient conditions to construct father and mother wavelets in the Fourier domain. Their analogues in time-domain have the following form (recall that $m_0(\xi) = \frac{1}{\sqrt{2}} \sum_k h_k e^{-ik\xi}$).

LEMMA 5.4 *The mother wavelet satisfies*

$$\psi(x) = \sqrt{2} \sum_k \lambda_k \varphi(2x - k), \qquad\qquad (5.13)$$

where $\lambda_k = (-1)^{k+1} \bar{h}_{1-k}$. For the father wavelet

$$\varphi(x) = \sqrt{2} \sum_k h_k \varphi(2x - k), \qquad\qquad (5.14)$$

we have the relations

$$\sum_k \bar{h}_k h_{k+2l} = \delta_{0l}, \\ \frac{1}{\sqrt{2}} \sum_k h_k = 1. \qquad\qquad (5.15)$$

Proof We have

$$\overline{m_0\left(\frac{\xi}{2} + \pi\right)} = \overline{\frac{1}{\sqrt{2}} \sum_k h_k e^{-ik(\xi/2 + \pi)}}$$

$$= \frac{1}{\sqrt{2}} \sum_k \bar{h}_k e^{ik(\xi/2 + \pi)} = \frac{1}{\sqrt{2}} \sum_k \bar{h}_k (-1)^k e^{ik\xi/2}.$$

Hence, by (5.9)

$$\hat{\psi}(\xi) = \sqrt{2} \sum_k \bar{h}_k (-1)^k e^{i(k-1)\xi/2} \frac{1}{2} \hat{\varphi}\left(\frac{\xi}{2}\right)$$

$$= \sqrt{2} \sum_{k'} \bar{h}_{1-k'} (-1)^{k'+1} e^{-ik'\xi/2} \frac{1}{2} \hat{\varphi}\left(\frac{\xi}{2}\right) \qquad (k' = 1 - k)$$

$$= \sqrt{2} \sum_k \lambda_k e^{-ik\xi/2} \frac{1}{2} \hat{\varphi}\left(\frac{\xi}{2}\right).$$

Taking the inverse Fourier transform of both sides, we get (5.13).

We now prove the first relation in (5.15). It is the time-domain version of the equality in Lemma 5.2. In fact, the equality

$$m_0(\xi)\overline{m_0(\xi)} + m_0(\xi + \pi)\overline{m_0(\xi + \pi)} = 1,$$

reads as

$$
\begin{aligned}
1 &= \frac{1}{2}\sum_{k,k'}\bar{h}_k h_{k'} e^{-i\xi(k'-k)} + \frac{1}{2}\sum_{k,k'}\bar{h}_k h_{k'} e^{-i\xi(k'-k)-i(k'-k)\pi} \\
&= \frac{1}{2}\sum_{k,k'}\bar{h}_k h_{k'} e^{-i\xi(k'-k)}\left[1 + e^{-i(k'-k)\pi}\right] \\
&= \sum_{l=-\infty}^{\infty}\sum_{k}\bar{h}_k h_{k+2l} e^{-2i\xi l}.
\end{aligned}
$$

The second relation in (5.15) is straightforward since $m_0(0) = 1$ (cf. (5.12)). □

5.3 Additional remarks

REMARK 5.4 In some works on wavelets one finds (5.13) in a different form, with $\lambda_k = (-1)^k \bar{h}_{1-k}$, or with other definition of λ_k which can be obtained for a certain choice of a function $\theta(\xi)$ (see Remark 5.3). This again reflects the fact that the mother wavelet is not unique, given a father wavelet.

REMARK 5.5 From (5.12) we deduce $|m_0(\pi)|^2 = 1 - |m_0(0)|^2 = 0$. Hence,

$$m_0(\pi) = 0, \tag{5.16}$$

which, in view of (5.9), entails

$$\hat{\psi}(0) = 0. \tag{5.17}$$

In other words,

$$\int \psi(x)dx = 0. \tag{5.18}$$

Note that $\int \varphi(x)dx \neq 0$, and it is always possible to impose $\int \varphi(x)dx = 1$; this last condition is satisfied for all examples of wavelets considered below. More discussion on these conditions is provided in Chapter 8.

It is natural to ask the following reverse question: *How to construct fathers from a mother?* To be more precise, let ψ be an $L_2(I\!R)$ function such that

$$\{2^{j/2}\psi(2^j x - k),\ j \in Z\!\!Z,\ k \in Z\!\!Z\},$$

is an ONB of $L_2(I\!R)$. Is ψ the mother wavelet of a MRA? At this level of generality the answer is *no*. But under mild regularity conditions, as studied in Lemarié-Rieusset(1993, 1994) and Auscher (1992), the question can be answered positively.

Chapter 6

Construction of wavelet bases

In Chapter 5 we derived general conditions on the functions φ and ψ that guarantee the wavelet expansion (3.5). It was shown that to find an appropriate pair (φ, ψ) it suffices, in fact, to find a father wavelet φ. Then one can derive a mother wavelet ψ, given φ. In this chapter we discuss two concrete approaches to the construction of father wavelets. The first approach is starting from Riesz bases, and the second approach is starting from a function m_0. For more details on wavelet basis construction we refer to Daubechies (1992),Chui(1992a, 1992b), Meyer (1993), Young (1993), Cohen & Ryan (1995), Holschneider (1995), Kahane & Lemarié-Rieusset (1995), Kaiser (1995).

6.1 Construction starting from Riesz bases

DEFINITION 6.1 *Let* $g \in L_2(I\!R)$. *The system of functions* $\{g(\cdot - k),\ k \in Z\!\!\!Z\}$ *is called* **Riesz basis** *if there exist positive constants* A *and* B *such that for any finite set of integers* $\Lambda \subset Z\!\!\!Z$ *and real numbers* $\lambda_i, i \in \Lambda$, *we have*

$$A \sum_{i \in \Lambda} \lambda_i^2 \leq \|\sum_{i \in \Lambda} \lambda_i g(\cdot - i)\|_2^2 \leq B \sum_{i \in \Lambda} \lambda_i^2.$$

In words, for the function belonging to the space spanned by the Riesz basis $\{g(\cdot - k),\ k \in Z\!\!\!Z\}$ the L_2 norm is equivalent to the l_2 norm of the coefficients (i.e. the system behaves approximately as an orthonormal basis).

47

PROPOSITION 6.1 *Let* $g \in L_2(I\!R)$. *The system of functions* $\{g(\cdot - k),\ k \in Z\!\!\!Z\}$ *is a* **Riesz basis** *if and only if there exist* $A > 0,\ B > 0$ *such that*

$$A \leq \sum_k |\hat{g}(\xi + 2\pi k)|^2 \leq B \qquad (a.e.). \qquad (6.1)$$

In this case we call $g(\cdot)$ *the* **generator function** , *and we call*

$$\Gamma(\xi) = \left(\sum_k |\hat{g}(\xi + 2\pi k)|^2 \right)^{1/2}$$

the **overlap function** *of the Riesz basis.*

Proof Using the Plancherel formula and the fact that Γ is periodic we have

$$
\begin{aligned}
\int |\sum_{k \in \Lambda} \lambda_k g(x - k)|^2 dx &= \frac{1}{2\pi} \int |\sum_{k \in \Lambda} \lambda_k \hat{g}(\xi) e^{-ik\xi}|^2 d\xi \\
&= \frac{1}{2\pi} \int |\sum_{k \in \Lambda} \lambda_k e^{-ik\xi}|^2 |\hat{g}(\xi)|^2 d\xi \\
&= \frac{1}{2\pi} \sum_l \int_{2\pi l}^{2\pi(l+1)} |\sum_{k \in \Lambda} \lambda_k e^{-ik\xi}|^2 |\hat{g}(\xi)|^2 d\xi \\
&= \frac{1}{2\pi} \sum_l \int_0^{2\pi} |\sum_{k \in \Lambda} \lambda_k e^{-ik\xi}|^2 |\hat{g}(\xi + 2\pi l)|^2 d\xi \\
&= \frac{1}{2\pi} \int_0^{2\pi} |\sum_{k \in \Lambda} \lambda_k e^{-ik\xi}|^2 |\Gamma(\xi)|^2 d\xi.
\end{aligned}
$$

Then it is clear that if (6.1) holds, the function g generates a Riesz basis. The proof of the inverse statement is given in Appendix D.   □

The idea how to construct a father wavelet is the following. Pick a generator function $g(\cdot)$. It is not necessarily a father wavelet, since a Riesz basis is not necessarily an orthonormal system. But it is straightforward to orthonormalize a Riesz basis as follows.

LEMMA 6.1 *Let* $\{g(\cdot - k),\ k \in Z\!\!\!Z\}$ *be a Riesz basis, and let* $\varphi \in L_2(I\!R)$ *be a function defined by its Fourier transform*

$$\hat{\varphi}(\xi) = \frac{\hat{g}(\xi)}{\Gamma(\xi)},$$

where

$$\Gamma(\xi) = \left(\sum_k |\hat{g}(\xi + 2\pi k)|^2 \right)^{1/2}$$

is the overlap function of the Riesz basis. Then $\{\varphi(\cdot - k), \ k \in \mathbb{Z}\}$ *is ONS.*

Proof Use Parseval's identity (4.4) and the fact that the Fourier transform of $\varphi(x - k)$ is $e^{-ik\xi}\hat{\varphi}(\xi)$ (see (4.5)). This gives

$$\int \varphi(x - k)\overline{\varphi(x - l)}dx = \frac{1}{2\pi} \int e^{-i(k-l)\xi} |\hat{\varphi}(\xi)|^2 d\xi$$

$$= \frac{1}{2\pi} \int e^{-i(k-l)\xi} \frac{|\hat{g}(\xi)|^2}{\Gamma^2(\xi)} d\xi = \frac{1}{2\pi} \sum_{m=-\infty}^{\infty} \int_{2\pi m}^{2\pi(m+1)} \frac{e^{-i(k-l)\xi}}{\Gamma^2(\xi)} |\hat{g}(\xi)|^2 d\xi$$

$$= \frac{1}{2\pi} \int_0^{2\pi} \frac{e^{-i(k-l)\xi}}{\Gamma^2(\xi)} \sum_{m=-\infty}^{\infty} |\hat{g}(\xi + 2\pi m)|^2 d\xi$$

$$= \frac{1}{2\pi} \int_0^{2\pi} e^{-i(k-l)\xi} d\xi = \delta_{kl},$$

where we used the fact that $\Gamma(\xi)$ is 2π-periodic. $\qquad\qquad\Box$

EXAMPLE 6.1 *B-splines.* Set

$$g_1(x) = I\{x \in (0,1]\},$$

and consider the generator function

$$g_N = \underbrace{g_1 * g_1 * \ldots * g_1}_{N-\text{times}}.$$

The function g_N is called *B-spline*. Let $\delta f(x) = f(x) - f(x - 1)$. The N-th iteration is

$$\delta^N f(x) = \sum_{k=0}^{N} \binom{N}{k}(-1)^k f(x - k).$$

Then the generator function g_N is given by

$$\delta^N \left[I\{x > 0\} \frac{x^{N-1}}{(N-1)!} \right]. \tag{6.2}$$

This formula can be proved by recurrence. In fact, observe that the Fourier transform of g_N is

$$\hat{g}_N(\xi) \;=\; \left(e^{-i\xi/2}\frac{\sin(\xi/2)}{(\xi/2)}\right)^N \tag{6.3}$$

$$=\; \frac{1-e^{-i\xi}}{i\xi}\hat{g}_{N-1}(\xi).$$

Applying the inverse Fourier transform to the last expression and using (4.5), (4.10) we see that

$$\frac{d}{dx}g_N(x) = g_{N-1}(x) - g_{N-1}(x-1) = \delta g_{N-1}(x).$$

Hence

$$g_N(x) = \int_0^x \delta g_{N-1}(t)dt = \delta \int_0^x g_{N-1}(t)dt.$$

Observing that $g_1 = \delta I\{x > 0\}\frac{x^0}{0!}$ we arrive after $N-1$ iterations at (6.2). Clearly, *supp* g_N is of the length N. The first two functions g_N are shown in Figure 6.1.

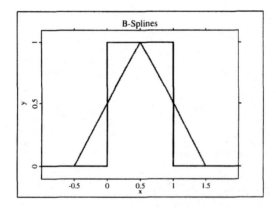

Figure 6.1: The first 2 elements of the B-spline Riesz basis.

If $N = 1$, then $g = g_1$ is the Haar father wavelet. The function g_2 is called *piecewise-linear B-spline*.

PROPOSITION 6.2 *The system* $\{g_N(\cdot - k),\ k \in \mathbb{Z}\}$, *for every* $N \geq 1$, *is a Riesz basis.*

Proof The Fourier transform of g_N is given in (6.3). The series

$$\sum_k |\hat{g}_N(\xi + 2\pi k)|^2$$

converges uniformly to some bounded function, since it is 2π-periodic, and for $\xi \in [0, 2\pi]$

$$|\hat{g}_N(\xi + 2\pi k)|^2 = \left| \frac{\sin\left(\frac{\xi}{2} + \pi k\right)}{\frac{\xi}{2} + \pi k} \right|^{2N} \leq \frac{1}{\left| \frac{\xi}{2} + \pi k \right|^{2N}} \leq \frac{1}{(\pi k)^{2N}}.$$

This entails for some $B > 0$ the condition

$$\sum_k |\hat{g}_N(\xi + 2\pi k)|^2 \leq B, \quad \forall \xi.$$

Now, since $\frac{\sin(x)}{x}$ is decreasing on $[0, \pi/2]$, we get (if $\xi \in [0, \pi]$)

$$\sum_k \left| \frac{\sin\left(\frac{\xi}{2} + \pi k\right)}{\frac{\xi}{2} + \pi k} \right|^{2N} \geq \left| \frac{\sin(\xi/2)}{(\xi/2)} \right|^{2N} \geq \left| \frac{\sin(\pi/4)}{\pi/4} \right|^{2N} = \left(\frac{2\sqrt{2}}{\pi} \right)^{2N}. \quad (6.4)$$

Quite similarly, for $\xi \in [\pi, 2\pi]$ we get the bound

$$\sum_k \left| \frac{\sin\left(\frac{\xi}{2} + \pi k\right)}{\frac{\xi}{2} + \pi k} \right|^{2N} \geq \left| \frac{\sin\left(\frac{\xi}{2} - \pi\right)}{\frac{\xi}{2} - \pi} \right|^{2N} = \left| \frac{\sin(\xi'/2)}{(\xi'/2)} \right|^{2N} \geq \left(\frac{2\sqrt{2}}{\pi} \right)^{2N},$$

where $\xi' = 2\pi - \xi \in [0, \pi]$, and we used the same argument as in (6.4). Thus, we proved the existence of $A > 0$ such that

$$\sum_k |\hat{g}_N(\xi + 2\pi k)|^2 \geq A.$$

Hence, (6.1) follows.   □

Let, for example, $N = 2$ (piecewise-linear B-spline generator function). Then

$$\hat{g}_2(\xi) = e^{-i\xi} \left(\frac{\sin(\xi/2)}{\xi/2} \right)^2,$$

and the sum $\Gamma^2(\xi)$ can be calculated explicitly (Daubechies (1992, Chap. 5)) :

$$\sum_k |\hat{g}(\xi + 2\pi k)|^2 = \sum_k \left| \frac{\sin\left(\frac{\xi}{2} + \pi k\right)}{\frac{\xi}{2} + \pi k} \right|^4 = \frac{2 + \cos \xi}{3}.$$

Hence, the father wavelet φ has the Fourier transform

$$\hat{\varphi}(\xi) = \sqrt{\frac{3}{2 + \cos \xi}} e^{-i\xi} \left(\frac{\sin(\xi/2)}{\xi/2} \right)^2.$$

It is called *Battle-Lemarié* father wavelet. How does the father wavelet look like? Let us denote by a_k the Fourier coefficients of the function $\sqrt{\frac{3}{2+\cos\xi}}$. These coefficients can be calculated numerically. Then

$$\sqrt{\frac{3}{2 + \cos \xi}} = \sum_k a_k e^{-ik\xi},$$

where an infinite number of a_k's are nonzero. Thus, $\hat{\varphi}(\xi)$ is an infinite sum

$$\hat{\varphi}(\xi) = \sum_k a_k e^{-i(k+1)\xi} \left(\frac{\sin(\xi/2)}{\xi/2} \right)^2,$$

and

$$\varphi(x) = \sum_k a_k g_2(x - k).$$

This father wavelet has the following properties:

· it is symmetric: $a_k = a_{-k}$, since $\sqrt{\frac{3}{2+\cos\xi}}$ is even,

· it is piecewise linear,

· *supp* $\varphi = \mathbb{R}$.

The Battle-Lemarié father wavelet is shown in Figure 6.2. Using the expression for $\hat{\varphi}$, we find now the function $m_0(\xi)$:

$$m_0(\xi) = \frac{\hat{\varphi}(2\xi)}{\hat{\varphi}(\xi)} = e^{-i\xi} \cos^2 \frac{\xi}{2} \sqrt{\frac{2 + \cos \xi}{2 + \cos 2\xi}}.$$

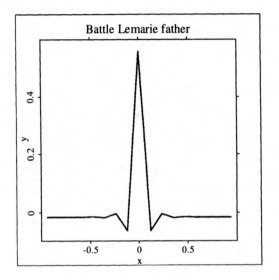

Figure 6.2: Battle-Lemarié father wavelet (N=2).

Then

$$m_1(\xi) = \overline{m_0(\xi + \pi)} e^{-i\xi} = \sin^2 \frac{\xi}{2} \sqrt{\frac{2 - \cos \xi}{2 + \cos 2\xi}},$$

and, by (5.4),

$$\hat{\psi}(\xi) = m_1\left(\frac{\xi}{2}\right) \hat{\varphi}\left(\frac{\xi}{2}\right) = \frac{\sin^4(\xi/4)}{(\xi/4)^2} \sqrt{\frac{2 - \cos \xi/2}{2 + \cos \xi}} \sqrt{\frac{3}{2 + \cos(\xi/2)}} e^{-i\xi/2}.$$

The inverse Fourier transform of this function gives the mother wavelet ψ. Again, one can calculate the Fourier coefficients of ψ only numerically. It is clear that

- $\psi(x)$ is symmetric around the point $x = 1/2$,

- ψ is piecewise-linear, since one can write

$$\psi(x) = \sum_k a'_k g_2(x - k),$$

where a'_k are some coefficients,

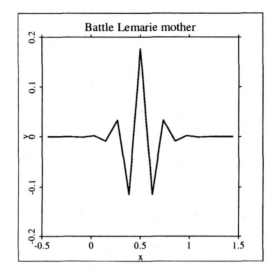

Figure 6.3: Battle-Lemarié mother wavelet (N=2).

· $supp \; \psi = I\!\!R$.

The Battle-Lemarié mother wavelet is shown in Figure 6.3. For $N > 2$ Battle-Lemarié wavelets are smoother, but they look in general similar to the case $N = 2$.

6.2   Construction starting from m_0

A disadvantage of the Riesz basis approach is that, except for the Haar case, one cannot find in this way compactly supported father and mother wavelets. Compactly supported wavelets are desirable from a numerical point of view. This is why we consider the second approach which allows to overcome this problem. Pick a function m_0 satisfying (5.12). By (5.2)

$$\hat{\varphi}(\xi) = m_0\left(\frac{\xi}{2}\right)\hat{\varphi}\left(\frac{\xi}{2}\right) = m_0\left(\frac{\xi}{2}\right)m_0\left(\frac{\xi}{4}\right)\hat{\varphi}\left(\frac{\xi}{4}\right) = \ldots$$

Continuing this splitting infinitely, and assuming that $\hat{\varphi}(0) = 1$ (see Section 5.2 and Remark 5.5), we arrive at the representation

$$\hat{\varphi}(\xi) = \prod_{j=1}^{\infty} m_0 \left(\frac{\xi}{2^j}\right) \tag{6.5}$$

provided the infinite product converges. Thus we could construct the father wavelet. However, this rises several questions.

Question 6.1: *When does the infinite product (6.5) converge pointwisely?*

Question 6.2: *If this product converges, does φ belong to $L_2(\mathbb{R})$?*

Question 6.3: *If φ is constructed in this way, is $\{\varphi(\cdot - k), \ k \in \mathbb{Z}\}$ an ONS?*

The following lemma answers Question 6.1.

LEMMA 6.2 *If $m_0(\xi)$ is Lipschitz continuous, then the infinite product in (6.5) converges uniformly on any compact set in $\mathbb{R}$.*

Proof. Since $m_0(0) = 1$,

$$\prod_{j=1}^{\infty} m_0 \left(\frac{\xi}{2^j}\right) = \prod_{j=1}^{\infty} \left(1 + u\left(\frac{\xi}{2^j}\right)\right),$$

where

$$\left| u\left(\frac{\xi}{2^j}\right) \right| = \left| m_0\left(\frac{\xi}{2^j}\right) - m_0(0) \right| \leq \frac{LK}{2^j}, \quad |\xi| \leq K.$$

Here L is the Lipschitz constant and $K > 0$ is arbitrary. Hence, the infinite product converges uniformly on every compact set of ξ's. $\square$

The examples of $m_0(\xi)$ used for the construction of $\hat{\varphi}(\xi)$ in practice are all of the form of trigonometric polynomials, that is

$$m_0(\xi) = \frac{1}{\sqrt{2}} \sum_{k=N_0}^{N_1} h_k e^{-ik\xi} \tag{6.6}$$

where $N_0, N_1 \in \mathbb{Z}$ are fixed, and

$$\frac{1}{\sqrt{2}} \sum_{k=N_0}^{N_1} h_k = 1, \qquad (\Longleftrightarrow m_0(0) = 1). \tag{6.7}$$

For this choice of m_0 the conditions of Lemma 6.2 are obviously satisfied. Moreover, the following result holds, answering Questions 6.2 and 6.3.

LEMMA 6.3 *Let m_0 be of the form (6.6), satisfying (6.7) and*

$$|m_0(\xi)|^2 + |m_0(\xi + \pi)|^2 = 1. \tag{6.8}$$

Assume also that there exists a compact set $\mathcal{K}$ in $I\!\!R$, containing a neighborhood of 0, such that

(1) $\sum_k I\{\xi + 2k\pi \in \mathcal{K}\} = 1$ (a.e.),

(2) $m_0(2^{-j}\xi) \neq 0$, $\forall\, \xi \in \mathcal{K}$, $\forall\, j \geq 1$.

Then the function $\hat{\varphi}(\xi)$ in (6.5) is the Fourier transform of a function $\varphi \in L_2(I\!\!R)$ such that

(i) supp $\varphi \subseteq [N_0, N_1]$,

and

(ii) $\{\varphi(\cdot - k),\ k \in \mathbb{Z}\}$ is an ONS in $L_2(I\!\!R)$.

This Lemma is due to Cohen. For the proof see Cohen & Ryan (1995) or Daubechies (1992, Chap. 6).   □

REMARK 6.1 The conditions (1) and (2) of Lemma 6.3 are obviously fulfilled if $\mathcal{K} = [-\pi, \pi]$ and $m_0(\xi) \neq 0$ for $|\xi| \leq \frac{\pi}{2}$.

Note that condition (6.8), in view of (6.6), may be written in terms of $\{h_{N_0}, \ldots, h_{N_1}\}$. Thus, we have only 2 restrictions, (6.7) and (6.6), on $N_1 - N_0 + 1$ coefficients. If $N_1 - N_0 + 1 > 2$, then there exist many possible solutions $\hat{\varphi}$, all giving father wavelets.

How to choose $\{h_k\}_{k=N_0,\ldots,N_1}$? First, note that every solution φ has compact support in view of Lemma 6.3 (i). This is a computational advantage with respect to the Riesz basis approach. Another advantage is that one can choose $\{h_k\}$ so that the father wavelet φ as well as the mother wavelet ψ:

· have a prescribed number of vanishing moments,

· have a prescribed number of continuous derivatives.

Note that the number of vanishing moments is linked to the rate of approximation of the wavelet expansion as will be shown in Chapter 8. This is the reason why it is important to be controlled.

Let us discuss the conditions on $\{h_k\}$, guaranteeing a prescribed number of vanishing moments. Consider first the father wavelets.

LEMMA 6.4 *Let the conditions of Lemma 6.3 be satisfied, and let*

$$\sum_{k=N_0}^{N_1} h_k k^l = 0, \qquad l = 1, \ldots, n. \tag{6.9}$$

Then for φ defined as the inverse Fourier transform of (6.5) we have

$$\int \varphi(x) x^l \, dx = 0, \qquad l = 1, \ldots, n. \tag{6.10}$$

Proof Condition (6.9) implies in view of the definition of $m_0(\xi)$ in (6.6):

$$m_0^{(l)}(0) = 0, \qquad l = 1, \ldots, n.$$

Since for any $\hat{\varphi}$ satisfying (6.5) we have

$$\hat{\varphi}(\xi) = \hat{\varphi}\left(\frac{\xi}{2}\right) m_0\left(\frac{\xi}{2}\right),$$

therefore also

$$\hat{\varphi}^{(l)}(0) = 0, \qquad l = 1, \ldots, n. \tag{6.11}$$

Note that $\hat{\varphi}(\xi)$ is n times continuously differentiable at $\xi = 0$, which follows from the fact that $\varphi \in L_2(I\!\!R)$ and $\varphi(x)$ is compactly supported (cf. (4.9)). Now, (6.10) is just a rewriting of (6.11). $\qquad\square$

Consider mother wavelets now. That is, take the function ψ which is the inverse Fourier transform of

$$\hat{\psi}(\xi) = \overline{m_0\left(\frac{\xi}{2} + \pi\right)} e^{-i\xi/2} \hat{\varphi}(\xi/2)$$

where $\hat{\varphi}(\xi)$ is defined by (6.5), or, in time domain (cf. Lemma 5.4):

$$\psi(x) = \sqrt{2} \sum_k \lambda_k \varphi(2x - k), \qquad \lambda_k = (-1)^{k+1} \bar{h}_{1-k}. \tag{6.12}$$

LEMMA 6.5 *Let the conditions of Lemma 6.3 be satisfied. Then $\psi \in L_2(I\!\!R)$, ψ is compactly supported, and*

$$supp \, \psi \subseteq \left[\frac{1}{2}(1 - N_1 + N_0), \frac{1}{2}(1 - N_0 + N_1)\right] \tag{6.13}$$

If, in addition,

$$\sum_{k} \lambda_k k^l = \sum_{k=1-N_1}^{1-N_0} (-1)^k \bar{h}_k (1-k)^l = 0, \qquad l = 1, \ldots, n, \qquad (6.14)$$

then

$$\int \psi(x) x^l \, dx = 0, \qquad l = 1, \ldots, n. \qquad (6.15)$$

Proof First, $\psi \in L_2(\mathbb{R})$, since we have $\varphi \in L_2(\mathbb{R})$ (Lemma 6.3), (6.12) and the definition of $m_0(\xi)$. To prove (6.13) note that in (6.12) we have only a finite number of summands such that:

$$\begin{aligned} N_0 \leq 1 - k \leq N_1 \quad &(\text{only these } \lambda_k \neq 0), \\ N_0 \leq 2x - k \leq N_1 \quad &(supp\ \varphi \subseteq [N_0, N_1]). \end{aligned} \qquad (6.16)$$

From (6.16) one gets:

$$1 - N_1 + N_0 \leq 2x \leq 1 - N_0 + N_1,$$

which gives (6.13).

Let us show (6.15). The equalities (6.15) are equivalent to:

$$\hat{\psi}^{(l)}(0) = 0, \qquad l = 1, \ldots, n. \qquad (6.17)$$

Now,

$$\hat{\psi}(\xi) = m_1 \left(\frac{\xi}{2} \right) \hat{\varphi} \left(\frac{\xi}{2} \right), \qquad (6.18)$$

where

$$m_1(\xi) = \overline{m_0(\xi + \pi)} e^{-i\xi} = \frac{1}{\sqrt{2}} \sum_{k} \lambda_k e^{-ik\xi},$$

and (6.14) entails:

$$m_1^{(l)}(0) = 0, \qquad l = 1, \ldots, n. \qquad (6.19)$$

Using this and (6.18) one arrives at (6.17).   □

REMARK 6.2 Clearly, (6.14) can be satisfied only if $n + 1$ is smaller than the degree of the polynomial $m_0(\xi)$, since (6.14) contains n equalities, and one has also the equality (6.7) on the coefficients of $m_0(\xi)$.

The problem of providing a prescribed number of continuous derivatives of φ and ψ is solved in a similar way: one should guarantee the existence of certain moments of $\hat{\varphi}(\xi)$ and $\hat{\psi}(\xi)$.

Chapter 7

Compactly supported wavelets

7.1   Daubechies' construction

The original construction of compactly supported wavelets is due to Daubechies (1988). Here we sketch the main points of Daubechies' theory. We are interested to find the exact form of functions $m_0(\xi)$, which are trigonometric polynomials, and produce father φ and mother ψ with compact supports such that, in addition, the moments of φ and ψ of order from 1 to n vanish. This property is necessary to guarantee good approximation properties of the corresponding wavelet expansions, see Chapter 8.

We have seen that the conditions of Lemma 6.3, together with (6.9) and (6.14) are sufficient for these purposes. So, we will assume that these conditions are satisfied in this section. An immediate consequence of (6.14) is the following

COROLLARY 7.1 *Assume the conditions of Lemma 6.3 and (6.14). Then $m_0(\xi)$ factorizes as*

$$m_0(\xi) = \left(\frac{1 + e^{-i\xi}}{2}\right)^{n+1} \mathcal{L}(\xi), \qquad (7.1)$$

where $\mathcal{L}(\xi)$ is a trigonometric polynomial.

Proof The relation (6.14) implies (6.19) which, in view of the definition of $m_1(\xi)$ is equivalent to

$$m_0^{(l)}(\pi) = 0, \qquad l = 1, \ldots, n.$$

Also $m_0(\pi) = 0$. Hence $m_0(\xi)$ has a zero of order $n + 1$ at $\xi = \pi$. This is exactly stated by (7.1). Since m_0 is a trigonometric polynomial, $\mathcal{L}(\xi)$ is also a trigonometric polynomial.   □

Corollary 7.1 suggests to look for functions $m_0(\xi)$ of the form

$$m_0(\xi) = \left(\frac{1 + e^{-i\xi}}{2}\right)^N \mathcal{L}(\xi), \tag{7.2}$$

where $N \geq 1$, and $\mathcal{L}(\xi)$ is a trigonometric polynomial. So we only need to find $\mathcal{L}(\xi)$. Denote

$$M_0(\xi) = |m_0(\xi)|^2.$$

Clearly $M_0(\xi)$ is a polynomial of $\cos \xi$ if $m_0(\xi)$ is a trigonometric polynomial. If, in particular, $m_0(\xi)$ satisfies (7.2), then

$$M_0(\xi) = \left(\cos^2 \frac{\xi}{2}\right)^N Q(\xi)$$

where $Q(\xi)$ is a polynomial in $\cos \xi$. Since $\sin^2 \frac{\xi}{2} = \frac{1 - \cos \xi}{2}$, we can write $Q(\xi)$ as a polynomial in $\sin^2 \frac{\xi}{2}$. Thus,

$$M_0(\xi) = \left(\cos^2 \frac{\xi}{2}\right)^N P\left(\sin^2 \frac{\xi}{2}\right),$$

where $P(\cdot)$ is a polynomial. In terms of P the constraint

$$\begin{aligned} |m_0(\xi)|^2 + |m_0(\xi + \pi)|^2 &= 1, \\ (\text{or} \quad M_0(\xi) + M_0(\xi + \pi) &= 1) \end{aligned}$$

becomes

$$(1 - y)^N P(y) + y^N P(1 - y) = 1, \tag{7.3}$$

which should hold for all $y \in [0, 1]$, and hence for all $y \in \mathbb{R}$.

Daubechies (1992, Chap. 6) gives the necessary and sufficient conditions on $P(\cdot)$ to satisfy (7.3). She shows that every solution of (7.3) is of the form

$$P(y) = \sum_{k=0}^{N-1} C_{N-1+k}^k y^k + y^N R(1/2 - y), \tag{7.4}$$

where $R(\cdot)$ is an odd polynomial such that $R(y) \geq 0$, $\forall\, y \in [0, 1]$.

Now, the function $\mathcal{L}(\xi)$, that we are looking for, is the "square root" of $P(\sin^2 \frac{\xi}{2})$, i.e. $|\mathcal{L}(\xi)|^2 = P(\sin^2 \frac{\xi}{2})$. Daubechies (1988) proposed to take in (7.4) $R \equiv 0$, and she showed that in this case $m_0(\xi)$ is such that

$$|m_0(\xi)|^2 = c_N \int_\xi^\pi \sin^{2N-1} x \, dx \qquad (7.5)$$

where the constant c_N is chosen so that $m_0(0) = 1$. For such functions $m_0(\xi)$ one can tabulate the corresponding coefficients h_k, see Daubechies (1992) and Table 1 in appendix A.

DEFINITION 7.1 *Wavelets constructed with the use of functions $m_0(\xi)$ satisfying (7.5) are called* **Daubechies wavelets.** *(One denotes them as D2N or Db2N.)*

EXAMPLE 7.1 Let $N = 1$. Then we obtain $D2$ wavelets. In this case $c_N = \frac{1}{2}$,

$$|m_0(\xi)|^2 = \frac{1}{2} \int_\xi^\pi \sin x \, dx = \frac{1 + \cos \xi}{2}.$$

Choose $m_0(\xi) = \frac{1+e^{-i\xi}}{2}$. Then

$$|m_0(\xi)|^2 = m_0(\xi)m_0(-\xi) = \frac{1 + \cos \xi}{2},$$

so this is the correct choice of $m_0(\xi)$. The function $\hat{\varphi}$ is computed easily. We have

$$\hat{\varphi}(\xi) = \lim_{n \to \infty} \prod_{j=1}^n \frac{1}{2}\left(1 + \exp\left(-\frac{i\xi}{2^j}\right)\right).$$

But

$$\prod_{j=1}^n \left(\frac{1 + e^{-i\xi/2^j}}{2}\right) = \prod_{j=1}^n \left(\frac{1 - e^{-i\xi/2^{j-1}}}{2(1 - e^{-i\xi/2^j})}\right)$$

$$= \frac{1}{2^n}\left(\frac{1 - e^{-i\xi}}{1 - e^{-i\xi/2^n}}\right) \xrightarrow[n \to \infty]{} \frac{1 - e^{-i\xi}}{i\xi}.$$

Hence

$$\hat{\varphi}(\xi) = \frac{1 - e^{-i\xi}}{i\xi}.$$

This implies that $\varphi(x)$ is the Haar father wavelet $\varphi(x) = I\{x \in (0, 1]\}$. *Thus, the Daubechies D2 wavelet coincides with the Haar wavelet.*

EXAMPLE 7.2 Let $N = 2$. Consider the $D4$ wavelet. One shows easily that $|m_0(\xi)|^2$ has the form

$$|m_0(\xi)|^2 = \frac{1}{4}(1 + \cos \xi)^2(2 - \cos \xi),$$

and the corresponding function $m_0(\xi)$ has the form

$$m_0(\xi) = \left(\frac{1 + e^{-i\xi}}{2}\right)^2 \frac{1 + \sqrt{3} + (1 - \sqrt{3})e^{-i\xi}}{2}.$$

In terms of coefficients h_k one has

$$m_0(\xi) = \frac{1}{\sqrt{2}} \sum_{k=0}^{3} h_k e^{-ik\xi}$$

where

$$
\begin{array}{ll}
h_0 = \frac{1+\sqrt{3}}{4\sqrt{2}}, & h_1 = \frac{3+\sqrt{3}}{4\sqrt{2}}, \\
h_2 = \frac{3-\sqrt{3}}{4\sqrt{2}}, & h_3 = \frac{1-\sqrt{3}}{4\sqrt{2}},
\end{array}
\tag{7.6}
$$

In general, for $N \geq 3$, the function $m_0(\xi)$ for $D2N$ has the form

$$
\begin{aligned}
m_0(\xi) &= \left(\frac{1 + e^{-i\xi}}{2}\right)^N \sum_{k=0}^{N-1} q_k e^{-ik\xi} \\
&= \frac{1}{\sqrt{2}} \sum_{k=0}^{2N-1} h_k e^{-ik\xi},
\end{aligned}
$$

where q_k are some coefficients.

REMARK 7.1 Properties of Daubechies' wavelets
By Lemma 6.3 (i) we have

$$supp\ \varphi \subseteq [0, 2N - 1] \tag{7.7}$$

and by (6.13)

$$supp\ \psi \subseteq [-N + 1, N]. \tag{7.8}$$

Since $m_0^{(l)}(\pi) = 0$, $l = 0, \ldots, N - 1$, we have

$$\int \psi(x)x^l\ dx = 0, \qquad l = 0, \ldots, N - 1. \tag{7.9}$$

The $D4$ wavelet for example satisfies $\int \psi(x)\, dx = 0, \quad \int x\, \psi(x)\, dx = 0.$

The Haar wavelet is the only symmetric compactly supported father wavelet, see Daubechies (1992).

We have the following smoothness property: for $N \geq 2$ the $D2N$ wavelets satisfy

$$\varphi, \psi \in H^{\lambda N}, \quad 0.1936 \leq \lambda \leq 0.2075, \tag{7.10}$$

where H^{λ} is the Hölder smoothness class with parameter λ. Asymptotically $\lambda = 0.2$, as $N \to \infty$.

EXAMPLE 7.3 As an example for this smoothness property consider the D4 wavelet. It is only 0.38-Hölderian, as (7.10) suggests.

Daubechies' wavelets are given in Figure 7.1. In this figure we show the father and the mother wavelets from $D2$(Haar) up to $D8$.

7.2 Coiflets

Daubechies' wavelets have vanishing moments for mother wavelets, but not for father wavelets. If the father wavelets have vanishing moments, the wavelet coefficients may be approximated by evaluations of the function f at discrete points: $\alpha_{jk} = 2^{-j/2} f\left(\frac{k}{2^j}\right) + r_{jk}$, with r_{jk} small enough. It can be a useful property in specific applications, see Section 3.3. Beylkin, Coifman & Rokhlin (1991) proposed a new class of wavelets which have essentially all the nice properties of Daubechies' wavelets and, in addition, vanishing moments of father wavelets. This class of wavelets (called *coiflets*) is discussed below.

To construct coiflets, one looks for $m_0(\xi)$ of the form

$$m_0(\xi) = \left(\frac{1 + e^{-i\xi}}{2}\right)^N \mathcal{L}(\xi),$$

where $\mathcal{L}(\xi)$ is a trigonometric polynomial. We want the following conditions to be satisfied

$$\begin{aligned}
&\int \varphi(x)\, dx = 1, \quad \int x^l \varphi(x)\, dx = 0, \quad l = 1, \ldots, N-1, \\
&\int \psi(x) x^l\, dx = 0, \quad\quad\quad\quad\quad\quad\quad l = 0, \ldots, N-1.
\end{aligned} \tag{7.11}$$

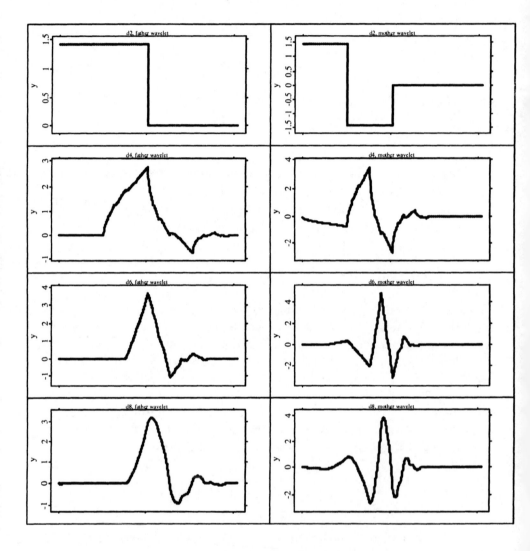

Figure 7.1: Daubechies' wavelets $D2$–$D8$.

These are equivalent to

$$\begin{cases} \hat{\varphi}(0) = 1, & \hat{\varphi}^{(l)}(0) = 0, \quad l = 1, \ldots, N-1, \\ \hat{\psi}^{(l)}(0) = 0, & \qquad\qquad\quad l = 0, \ldots, N-1. \end{cases}$$

The conditions $\hat{\varphi}^{(l)}(0) = 0$ are implied by (see the proof of Lemma 6.4)

$$\hat{m}_0^{(l)}(0) = 0, \qquad l = 1, \ldots, N-1. \tag{7.12}$$

COROLLARY 7.2 *Assume the conditions of Lemma 6.3 and (7.12). Then $m_0(\xi)$ can be represented as*

$$m_0(\xi) = 1 + (1 - e^{-i\xi})^N S(\xi) \tag{7.13}$$

where $S(\xi)$ is a trigonometric polynomial.

Proof follows the proof of Corollary 7.1.   □

Set $N = 2K$, K integer. Daubechies (1992, Chap. 8) shows, that (7.1) and (7.13) imply the following form of $m_0(\xi)$

$$m_0(\xi) = \left(\frac{1 + e^{-i\xi}}{2}\right)^{2K} P_1(\xi), \tag{7.14}$$

where

$$P_1(\xi) = \sum_{k=0}^{K-1} C_{K-1+k}^k \left(\sin^2 \frac{\xi}{2}\right)^k + \left(\sin^2 \frac{\xi}{2}\right)^K F(\xi)$$

and $F(\xi)$ is a trigonometric polynomial chosen so that

$$|m_0(\xi)|^2 + |m_0(\xi + \pi)|^2 = 1.$$

DEFINITION 7.2 *Wavelets obtained with the function $m_0(\xi)$ given in (7.14) are called* **coiflets** *(of order K), and denoted by CK (for example, $C1$, $C2$ etc.).*

REMARK 7.2 Properties of coiflets of order K.

$$\text{supp } \varphi \subseteq [-2K, 4K - 1]. \tag{7.15}$$

$$\text{supp } \psi \subseteq [-4K + 1, 2K]. \tag{7.16}$$

$$\int x^l \varphi(x) \, dx = 0, \qquad l = 1, \ldots, 2K - 1. \tag{7.17}$$

$$\int x^l \psi(x) \, dx = 0, \qquad l = 0, \ldots, 2K - 1. \tag{7.18}$$

$$\textit{Coiflets are not symmetric.} \tag{7.19}$$

EXAMPLE 7.4 As an example let us consider the $C3$ coiflet which has 5 vanishing moments,

$$supp\ \varphi_3 = [-6, 11],$$
$$supp\ \psi_3 = [-11, 6].$$

The coefficients $\{h_k\}$ for coiflets are tabulated in Daubechies (1992) and in Table 1 of appendix A. Examples of coiflets are given in Figure 7.2 where we show the father and mother coiflets $C1$ to $C4$. In the upper left we have plotted $C1$ and below $C2$. In the upper right we have father and mother of $C3$.

7.3   Symmlets

It is shown in Daubechies (1992) that except for the Haar system no system φ, ψ can be at the same time compactly supported and symmetric. Nevertheless, for practical purposes (in image processing for example), one can try to be as close as possible to the symmetry by requiring the following: the phase of $m_0(\xi)$ is minimal among all the $m_0(\xi)$ with the same value $|m_0(\xi)|$. This defines a certain choice of the polynomial $\mathcal{L}(\xi)$, with the least possible shift.

Coefficients $\{h_k\}$ for symmlets are tabulated in Daubechies (1992, p. 198). One uses the notation SN for symmlet of order N, (for example, $S1$, $S2$ etc.).

REMARK 7.3 Properties of symmlets. The symmlet SN has the father and mother wavelets such that

$$supp\ \varphi \subseteq [0, 2N - 1]. \tag{7.20}$$

$$supp\ \psi \subseteq [-N + 1, N]. \tag{7.21}$$

$$\int x^l \psi(x)\, dx = 0, \qquad l = 0, \ldots, N - 1. \tag{7.22}$$

$$\textit{Symmlets are not symmetric.} \tag{7.23}$$

EXAMPLE 7.5 The symmlet S8 has 7 vanishing moments (for mother wavelet only) and

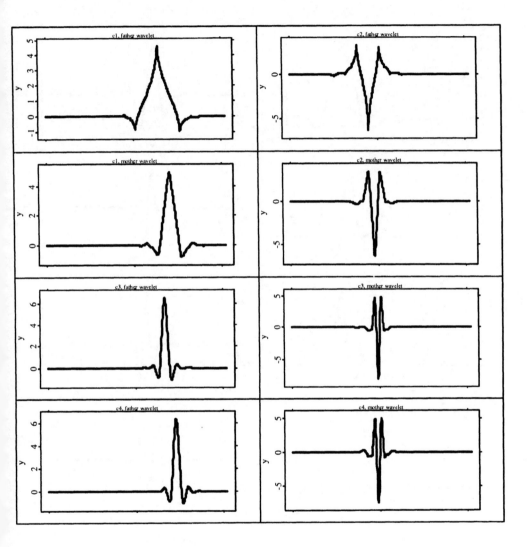

Figure 7.2: Coiflets in order $C1$ to $C4$.

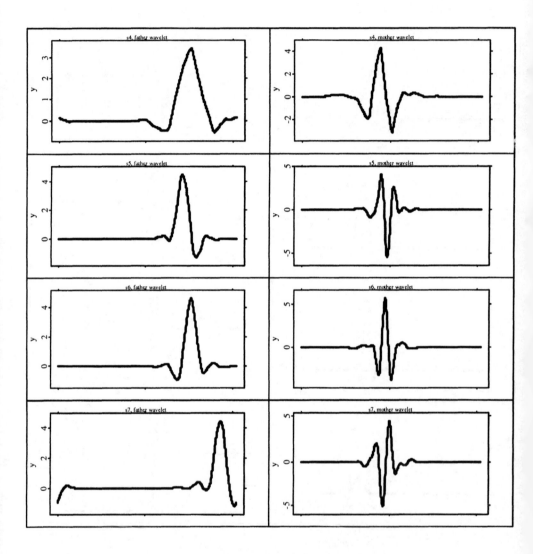

Figure 7.3: Four symmlets $S4$–$S7$.

$$supp\ \varphi_8 = [0, 15],$$
$$supp\ \psi_8 = [-7, 8].$$

The first four symmlets are shown in Figure 7.3.

Chapter 8

Wavelets and Approximation

8.1 Introduction

In this chapter we study the approximation properties of wavelet expansions on the Sobolev spaces. We specify how fast does the wavelet expansion converge to the true function f, if f belongs to some Sobolev space. This study is continued in Chapter 9 where we consider the approximation on the Besov spaces and show that it has an intrinsic relation to wavelet expansions. The presentation in this chapter and in Chapter 9 is more formal than in the previous ones. It is designed for the mathematically oriented reader who is interested in a deeper theoretical insight into the properties of wavelet bases.

We start by considering a general kernel approximation of functions on the Sobolev spaces. We give an approximation theorem: if f is in a Sobolev space and if the kernel satisfies a certain moment condition, then the approximation has a given accuracy. The theorem also admits an inverse (for periodic kernels): if the approximation is of the given accuracy at least for one function, then the kernel has to satisfy the moment condition. This main *moment condition* which requires that certain moments of the kernel were zero, is therefore in the focus of our study.

First, we restrict the class of kernels by the periodic projection kernels of the form $K(x,y) = \sum_k \varphi(x-k)\overline{\varphi(y-k)}$, where $\varphi \in L_2(I\!R)$ is such that $\{\varphi(x-k),\ k \in \mathbb{Z}\}$ is an orthonormal system. For these kernels the moment condition is essentially equivalent to good approximation properties. Therefore, we specify the assumptions on φ that ensure the moment condition for

71

such kernels.

Next, we restrict the class of kernels even more by assuming that φ is the scaling function of a multiresolution analysis (i.e. a father wavelet). We derive necessary and sufficient conditions for the moment condition in this case (Theorem 8.3) and provide the approximation theorem for wavelet expansions on the Sobolev spaces (Corollary 8.2). These are the main results of the chapter. Moreover, in Proposition 8.6 and Corollary 8.1 we prove that, under a mild condition on the father wavelet φ (for example, for any bounded and compactly supported father wavelet), the set $\bigcup_{j \geq 0} V_j$ is dense in $L_2(I\!R)$, and that certain other properties of MRA stated without proof in Chapters 3 and 5 are satisfied.

8.2   Sobolev Spaces

Let us first recall the definition of weak differentiability. Denote $\mathcal{D}(I\!R)$ the space of infinitely many times differentiable compactly supported functions. The following result is well known.

PROPOSITION 8.1 *Let f be a function defined on the real line which is integrable on every bounded interval. The two following facts are equivalent:*

1. *There exists a function g defined on the real line which is integrable on every bounded interval such that*

$$\forall x \leq y, \int_x^y g(u)du = f(y) - f(x)$$

2. *There exists a function g defined on the real line which is integrable on every bounded interval such that :*

$$\forall \phi \in D(I\!R) : \int f(u)\phi'(u)du = -\int g(u)\phi(u)du.$$

DEFINITION 8.1 *A function f satisfying the properties of Proposition 8.1 is called weakly differentiable. The function g is defined almost everywhere, is called the weak derivative of f and will be denoted by f'.*

It follows that any weakly differentiable function is continuous.

PROPOSITION 8.2 *Let f and g be weakly differentiable functions. Then fg is weakly differentiable, and $(fg)' = f'g + fg'$.*

Proof

Let $a \leq b$. By the Fubini theorem we have :

$$\{f(b) - f(a)\}\{g(b) - g(a)\} = \int_a^b f'(x)dx \int_a^b g'(y)dy = \int_a^b \int_a^b f'(x)g'(y)dxdy$$

We divide the domain of integration in two parts:

$$\int_a^b \int_a^b f'(x)g'(y)dxdy = \int_a^b f'(x) \int_a^x g'(v)dvdx + \int_a^b g'(y) \int_a^y f'(u)dudy.$$

Thus

$$\begin{aligned}
\{f(b) - f(a)\}\{g(b) - g(a)\} &= \int_a^b f'(x)\{g(x) - g(a)\}dx \\
&\quad + \int_a^b g'(y)\{f(y) - f(a)\}dy \\
&= \int_a^b \{f'(x)g(x) + g'(x)f(x)\}dx \\
&\quad -\{f(b) - f(a)\}g(a) - f(a)\{g(b) - g(a)\}.
\end{aligned}$$

Finally

$$\{f(b)g(b) - f(a)g(a)\} = \int_a^b (f'(x)g(x) + g'(x)f(x))dx$$

$\square$

DEFINITION 8.2 *A function f is N times weakly differentiable, if it has N-1 weakly differentiable weak derivatives. This implies that these derivatives $f, f', ...f^{(N-1)}$ are continuous.*

REMARK 8.1 If f has a weak derivative, we have for all x and y :

$$f(y) = f(x) + \int_0^1 f'(x + t(y - x))(y - x)dt.$$

If f is N times weakly differentiable, then, using recursively the integration by parts, one can easily prove the Taylor formula

$$f(y) = \sum_{k=0}^{N-1} \frac{f^{(k)}(x)}{k!}(y - x)^k + \int_0^1 (y - x)^N \frac{(1 - u)^{N-1}}{(N - 1)!} f^{(N)}(x + u(y - x))du.$$

Let us now define the Sobolev spaces. In the following we use the $L_p(I\!R)$ norms:

$$\|f\|_p = \begin{cases} (\int |f(x)|^p \, dx)^{1/p}, & \text{if } 1 \le p < \infty, \\ \operatorname{ess\,sup}_x |f(x)|, & \text{if } p = \infty. \end{cases}$$

DEFINITION 8.3 *Let* $1 \le p \le \infty, m \in \{0,1,\ldots\}$. *The function* $f \in L_p(I\!R)$ *belongs to the Sobolev space* $W_p^m(I\!R)$, *if it is m-times weakly differentiable, and if* $f^{(j)} \in L_p(I\!R)$, $j = 1,\ldots,m$. *In particular,* $W_p^0(I\!R) = L_p(I\!R)$.

It can be proved that in this definition it is enough to have $f^{(m)} \in L_p(I\!R)$ instead of $f^{(j)} \in L_p(I\!R)$, $j = 1,\ldots,m$.

The space $W_p^m(I\!R)$ is naturally equipped with the associated norm

$$\|f\|_{W_p^m} = \|f\|_p + \|f^{(m)}\|_p.$$

For the purpose of this section we define also the space $\tilde{W}_p^m(I\!R)$ which is very close to $W_p^m(I\!R)$.

DEFINITION 8.4 *The space* $\tilde{W}_p^m(I\!R)$ *is defined as follows. Set* $\tilde{W}_p^m(I\!R) = W_p^m(I\!R)$, *if* $1 \le p < \infty$, *and*

$$\tilde{W}_\infty^m(I\!R) = \{f \in W_\infty^m(I\!R) : f^{(m)} \text{is uniformly continuous}\}.$$

In particular, $\tilde{W}_p^0(I\!R) = L_p(I\!R)$, $1 \le p < \infty$.

Sometimes we write shortly W_p^m and $\tilde{W}_p^m$ instead of $W_p^m(I\!R)$ and $\tilde{W}_p^m(I\!R)$.

REMARK 8.2 Let $\tau_h f(x) = f(x-h)$, and define the modulus of continuity $\omega_p^1 f(t) = \sup_{|h| \le t} \|\tau_h f - f\|_p$. Then $f \in \tilde{W}_p^m(I\!R)$ if and only if the following two relations hold:

$$f \in W_p^m(I\!R) \tag{8.1}$$

and

$$\omega_p^1(f^{(m)}, t) \to 0, t \to 0. \tag{8.2}$$

In fact, $f \in L_p(I\!R)$ implies that f is continuous in $L_p(I\!R)$, for $1 \le p < \infty$.

For the general theory of Sobolev spaces see e.g. the books of Adams (1975), Bergh & Löfström (1976), Triebel (1992), DeVore & Lorentz (1993). We shall frequently use the following inequalities for the L_p-norms.

LEMMA 8.1 (Generalized Minkowsky inequality) *Let $f(x,y)$ be a Borel function on $I\!R \times I\!R$ and $1 \leq p \leq \infty$. Then*

$$\left\| \int_{I\!R} f(x,y)dx \right\|_p \leq \int_{I\!R} \|f(x,\cdot)\|_p dx.$$

LEMMA 8.2 *Let $f \in L_p(I\!R), g \in L_1(I\!R), 1 \leq p \leq \infty$. Then*

$$\|f * g\|_p \leq \|g\|_1 \|f\|_p.$$

Proof of these inequalities can be found in Adams (1975), Bergh & Löfström (1976), Triebel (1992), DeVore & Lorentz (1993). Note that Lemma 8.2 is an easy consequence of Lemma 8.1.

8.3 Approximation kernels

We develop here and later in this chapter the idea of Fix & Strang (1969).

DEFINITION 8.5 *A kernel $K(x,y)$ is a function defined on $I\!R \times I\!R$. If $K(x,y) = K(x-y)$, then K is called a convolution kernel.*

Let $K(x,y)$ be a kernel. For a positive real number h, define $K_h(x,y) = h^{-1}K(h^{-1}x, h^{-1}y)$. If $h = 2^{-j}$, we write $K_j(x,y)$ instead of $K_h(x,y)$. For a measurable function f we introduce the operator associated with the kernel: $K_h f(x) = \int K_h(x,y)f(y)dy$. Analogously, $K_j f$ and Kf are defined. The function $K_h f$ will play the role of an *approximation* for the function f, and we will evaluate how this approximation becomes close to f as h tends to 0.

Let us introduce some conditions on kernels used in the sequel. Let $N \geq 0$ be an integer.

Condition H *(size condition)* There exists an integrable function $F(x)$, such that $|K(x,y)| \leq F(x-y), \forall x, y \in I\!R$.

Condition $H(N)$ Condition H holds and $\int |x|^N F(x)dx < \infty$.

Condition P *(periodicity condition)* $K(x+1, y+1) = K(x,y), \forall x, y \in I\!R$.

Condition $M(N)$ *(moment condition) Condition H(N) is satisfied and*

$$\int K(x,y)(y-x)^k dy = \delta_{0k}, \quad \forall k = 0, \dots, N, \ \forall x \in \mathbb{R}, \qquad (8.3)$$

where δ_{jk} is the Kronecker delta.

REMARK 8.3 Condition H implies that for all h and for all $p, 1 \leq p \leq \infty$, we have

$$\|K_h f\|_p \leq \|F\|_1 \|f\|_p \qquad (8.4)$$

(cf. Lemmas 8.1 and 8.2). Condition P (periodicity) is obviously satisfied in the case of a convolution kernel $K(x,y) = K(x-y)$. The condition (8.3) is equivalent to the following one : $Kp = p$ for every polynomial p of degree not greater than N.

8.4   Approximation theorem in Sobolev spaces

Here we study the rates of convergence in L_p, as $h \to 0$, of the approximation $K_h f$ to the function f, when f belongs to a Sobolev space.

THEOREM 8.1 *Let K be a kernel, and let $N \geq 0$ be an integer.*

(i) *If K satisfies Condition $M(N)$ and if f belongs to the Sobolev space $\tilde{W}_p^N(\mathbb{R})$, then $h^{-N}\|K_h f - f\|_p \to 0$ when $h \to 0$, for any $p \in [1, \infty]$.*

(ii) *If K satisfies Conditions $M(N)$ and $H(N+1)$ and if f belongs to the Sobolev space $W_p^{N+1}(\mathbb{R})$, then $h^{-(N+1)}\|K_h f - f\|_p$ remains bounded when $h \to 0$, for any $p \in [1, \infty]$.*

(iii) *If K satisfies Conditions P and $H(N)$, if there exist $p \in [1, \infty]$ and a non constant function $f \in \tilde{W}_p^N(\mathbb{R})$, such that $h_n^{-N}\|K_{h_n} f - f\|_p \to 0$, for some positive sequence $h_n \to 0$, then K satisfies the condition $M(N)$.*

Proof Introduce the functions

$$\mu_0(x) \equiv \int K(x,y)dy - 1,$$

$$\mu_j(x) = \int K(x,y)\frac{(y-x)^j}{j!}dy, \quad j = 1, 2, \dots, N.$$

Observe that the functions $\mu_j(x)$ exist if K satisfies the Condition $H(N)$. Using the Taylor formula, we have for any f in the Sobolev space W_p^N :

$$f(y) = \sum_{k=0}^{N} \frac{f^{(k)}(x)}{k!}(y-x)^k + R_N f(y,x),$$

where $R_0 f(x,y) = f(y) - f(x)$,

$$R_N f(y,x) = \int_0^1 (y-x)^N \frac{(1-u)^{N-1}}{(N-1)!} \{f^{(N)}(x+u(y-x)) - f^{(N)}(x)\} du, \quad N \geq 1.$$

If moreover $f \in W_p^{N+1}$, then

$$R_N f(y,x) = \int_0^1 (y-x)^{N+1} \frac{(1-u)^N}{N!} f^{(N+1)}(x+u(y-x)) \, du.$$

Thus

$$K_h f(x) - f(x) = \sum_{k=0}^{N} \mu_k(h^{-1}x) f^{(k)}(x) h^k + \int K_h(x,y) R_N f(y,x) dy. \quad (8.5)$$

(i) Let K satisfy the Condition $M(N)$ and let $f \in \tilde{W}_p^N$. Then clearly, $\mu_j(x) = 0$ (a.e.), $j = 0, 1, \ldots, N$, and (8.5) yields

$$K_h f(x) - f(x) = \int K_h(x,y) R_N f(y,x) dy$$
$$= \int_0^1 du \int_{I\!R} K_h(x,y) \frac{(1-u)^{N-1}}{(N-1)!}(y-x)^N [f^{(N)}(x+u(y-x)) - f^{(N)}(x)] dy$$

and hence

$$|K_h f(x) - f(x)|$$
$$\leq h^N \int_0^1 du \int_{I\!R} |t|^N F(t) \frac{(1-u)^{N-1}}{(N-1)!} |f^{(N)}(x-tuh)) - f^{(N)}(x)| dt.$$

We used here the inequality $|K(x,y)| \leq F(x-y)$ and set $x - y = th$. Thus Lemma 8.1, Remark 8.2 and the fact that $f \in \tilde{W}_p^N$ give

$$\|K_h f - f\|_p \leq \frac{h^N}{(N-1)!} \int_0^1 (1-u)^{N-1} du \int_{I\!R} |t|^N F(t) \|\tau_{tuh}(f^{(N)}) - f^{(N)}\|_p o$$
$$= h^N o(h), \text{ as } h \to 0,$$

where $\tau_v f(x) = f(x-v), v \in I\!R$.

(ii) Let now $f \in W_p^{N+1}$. Then, as K satisfies Conditions $M(N)$ and $H(N+1)$, we have

$$K_h f(x) - f(x)$$
$$= \int_0^1 du \int_{I\!R} K_h(x,y) \frac{(1-u)^N}{N!} (y-x)^{N+1} f^{(N+1)}(x + u(y-x)) dy.$$

Thus

$$|K_h f(x) - f(x)|$$
$$\leq h^{N+1} \int_0^1 du \int_{I\!R} |t|^{N+1} F(t) \frac{(1-u)^N}{N!} |f^{(N+1)}(x + tuh))| dt,$$

and the application of Lemma 8.1 gives

$$\|K_h f - f\|_p \leq \frac{h^{N+1}}{N!} \int_0^1 du(1-u)^N \int_{I\!R} |t|^{N+1} F(t) \|f^{(N+1)}\|_p dt = O(h^{N+1}),$$

as $h \to 0$.

(iii) The periodicity condition on K implies that the functions $\mu_k(x)$, $k = 0, 1, \ldots, N$ are periodical, with period 1. By assumption, $\|K_h f - f\|_p = o(h_n^N)$. On the other hand, it follows from the proof of (i) that

$$\| \int K_h(x,y) R_l f(y,x) dy \|_p = o(h^l), \ l = 0, 1, \ldots, N.$$

This and (8.5) entail

$$\| \sum_{k=0}^l \mu_k(h_n^{-1} x) f^{(k)}(x) h_n^k \|_p = o(h_n^l).$$

Using Lemma 8.4, proved below, we get successively

$$\mu_0(x) = 0, \ \mu_1(x) = 0, \ \ldots, \mu_N(x) \equiv 0 \quad (a.e.).$$

The following two lemmas end the proof.   □

LEMMA 8.3 *(Adams (1975),Bergh & Löfström (1976), Triebel (1992)) Let θ be a bounded periodic function with period 1 and let $g \in L_1(I\!R)$.*

$$\int \theta(h^{-1}y)g(y) dy \to \int_0^1 \theta(u) du \int g(y) dy$$

as $h \to 0$.

Proof First consider the function g that is continuously differentiable and has support $\subset [a, b]$. We have

$$\int \theta(h^{-1}t)g(t)dt \;=\; h \int g(th)\theta(t)dt$$

$$=\; \sum_k h \int_0^1 g\{h(t+k)\}\theta(t)dt$$

$$=\; \int_0^1 \theta(t)S(t)dt,$$

where

$$S(t) = h \sum_k g(th + kh).$$

Clearly, $S(t)$ converges uniformly to $\int g(u)du$ for every t, as $h \to 0$. In fact,

$$\left| S(t) - \int_{-\infty}^{+\infty} g(th + u)du \right| = \left| \sum_m \int_{mh}^{(m+1)h} \{g(th + mh) - g(th + u)\}du \right|.$$

Note that, for $u \in [mh, (m+1)h]$,

$$|g(th + mh) - g(th + u)| \le h||g'||_\infty I\{t: \; a \le th + mh, th + (m+1)h \le b\}$$

and

$$\sum_m I\{t: \; a \le th + mh, th + (m+1)h \le b\} \le \frac{(L+1)}{h},$$

where L is the length of the support of g and I is the indicator function. Hence,

$$\left| S(t) - \int_{-\infty}^{+\infty} g(th + u)du \right| \le h||g'||_\infty (L+1),$$

which entails that $S(t)$ is uniformly bounded, if h is small. Applying the dominated convergence theorem, we get

$$\int_0^1 \theta(t)S(t)dt \;\to\; \int_0^1 \theta(u)du \int g(y)dy,$$

as $h \to 0$. For general functions g we use the fact that compactly supported differentiable functions are dense in $L_1(\mathbb{R})$. $\qquad\square$

LEMMA 8.4 *Let θ be a bounded periodic function with period 1 and let $h > 0$. If there exists a function $f \in L_p(I\!R)$ such that $f \neq 0$ and*

$$||\theta(h^{-1}x)f(x)||_p \to 0, \qquad (8.6)$$

as $h \to 0$, then $\theta = 0$ (a.e.).

Proof   Take a function $g \in L_q(I\!R)$, where $\frac{1}{p} + \frac{1}{q} = 1$, such that $\int fg \neq 0$. Denote by c_m the m-th Fourier coefficient of θ. Then, by Lemma 8.3

$$\int_{-\infty}^{\infty} \theta(h^{-1}t) \exp(-2\pi imh^{-1}t)f(t)g(t)dt \to c_m \int fg \qquad (8.7)$$

as $h \to 0$. The integral in the LHS of (8.7) does not exceed $||\theta(h^{-1}x)f(x)||_p \|g\|_q$ by the Hölder inequality. Hence, by assumption (8.6), this integral tends to 0, as $h \to 0$. This yields $c_m = 0$. Since m is arbitrary, this entails $\theta = 0$ (a.e.).   □

Parts (i) and (ii) of Theorem 8.1 indicate the rate of approximation of f by $K_h f$ provided that f is regular and K satisfies the moment condition $M(N)$. Part (iii) shows that the moment condition is crucial to guarantee the good approximation properties of $K_h f$. In Section 8.6 we shall investigate this condition further.

REMARK 8.4 If K satisfies the condition $M(0)$, then $\forall 1 \leq p < \infty$, $\forall f \in L_p(I\!R)$,

$$||K_j f - f||_p \to 0, \text{ as } j \to 0.$$

The same is true for $p = \infty$, if $f \in L_\infty(I\!R)$ and is uniformly continuous. This is due to the fact that $\tilde{W}_p^0 = L_p$, if $1 \leq p < \infty$, and that $\tilde{W}_\infty^0$ is the space of uniformly continuous bounded functions.

If $f \in L_\infty(I\!R)$, we have only a weak convergence of $K_j f$ to f in the following sense. For all $g \in L_1(I\!R)$, $\int g(x)K_j f(x)dx = \int f(u)\tilde{K}_j g(u)du$, where $\tilde{K}(u,v) = K(v,u)$. But this kernel satisfies also the condition $M(0)$, so by Theorem 8.1 (i) $||\tilde{K}_j g - g||_1 \to 0$. This implies:

$$\forall g \in L_1(I\!R), \int g(x)K_j f(x)dx \to \int f(x)g(x)dx, \text{ as } j \to \infty.$$

8.5 Periodic kernels and projection operators

DEFINITION 8.6 *A function $\varphi \in L_2(I\!R)$ such that $\{\varphi(x - k), \ k \in Z\!\!\!Z\}$ is an ONS, is called* **scaling function**.

For any function $f \in L_2(I\!R)$ its orthogonal projection operator P_{V_0} on V_0 is defined by

$$\int |P_{V_0}(f)(x) - f(x)|^2 dx = min_{g \in V_0} \int |g(x) - f(x)|^2 dx. \qquad (8.8)$$

Let $\varphi(\cdot)$ be a scaling function, let V_0 be the subspace of $L_2(I\!R)$ spanned by the orthogonal basis $\{\varphi(x - k), \ k \in Z\!\!\!Z\}$ and let $f \in L_2(I\!R)$. Then

$$P_{V_0}(f)(\cdot) = \sum_k (\int f(y)\overline{\varphi(y - k)}dy)\varphi(\cdot - k). \qquad (8.9)$$

The following condition on the scaling function φ will be useful in the sequel.

Condition (θ). *The function $\theta_\varphi(x) = \sum_k |\varphi(x - k)|$ is such that*

$$\operatorname{ess\,sup}_x \theta_\varphi(x) < \infty.$$

Note that if φ satisfies Condition (θ), then $\varphi \in L_\infty(I\!R)$, and also θ_φ is a periodic function with period 1, such that

$$\int_0^1 \theta_\varphi(x) \, dx < \infty. \qquad (8.10)$$

Also,

$$\int_{-\infty}^\infty |\varphi(x)|dx = \int_0^1 \sum_k |\varphi(x - k)|dx = \int_0^1 \theta_\varphi(x)dx < \infty. \qquad (8.11)$$

Hence, Condition (θ) implies that $\varphi \in L_1(I\!R) \cap L_\infty(I\!R)$, and thus the Fourier transform $\hat{\varphi}(\xi)$ is continuous, and $\varphi \in L_p(I\!R)$, $\forall 1 \le p \le \infty$.

Heuristically, Condition (θ) is a localization condition. Clearly, it holds for compactly supported bounded functions φ, and it is not satisfied for the Shannon function $\varphi(x) = \frac{\sin(\pi x)}{\pi x}$. It forbids the function φ to be too spread, for example, to have oscillations possibly accumulated in the sum over k.

The following proposition is a main tool for the evaluation of L_p-norms in the context of wavelets.

PROPOSITION 8.3 *If a function φ satisfies Condition (θ), then for any sequence $\{\lambda_k,\ k \in \mathbb{Z}\}$, satisfying $||\lambda||_{l_p} = (\sum_k |\lambda_k|^p)^{\frac{1}{p}} < \infty$, and any p and q such that $1 \leq p \leq \infty, \frac{1}{p} + \frac{1}{q} = 1$, we have:*

$$|| \sum_k \lambda_k \varphi(x - k)||_p \quad \leq \quad ||\lambda||_{l_p} ||\theta_\varphi||_\infty^{\frac{1}{q}} ||\varphi||_1^{\frac{1}{p}}, \tag{8.12}$$

$$|| \sum_k \lambda_k 2^{\frac{j}{2}} \varphi(2^j x - k)||_p \quad \leq \quad ||\lambda||_{l_p} 2^{(\frac{j}{2} - \frac{j}{p})} ||\theta_\varphi||_\infty^{\frac{1}{q}} ||\varphi||_1^{\frac{1}{p}}, \tag{8.13}$$

If, moreover, φ is a scaling function, then

$$C_1 ||\lambda||_{l_p} \leq || \sum_k \lambda_k \varphi(x - k)||_p \leq C_2 ||\lambda||_{l_p}, \tag{8.14}$$

$$C_1 ||\lambda||_{l_p} 2^{(\frac{j}{2} - \frac{j}{p})} \leq || \sum_k \lambda_k 2^{\frac{j}{2}} \varphi(2^j x - k)||_p \leq C_2 ||\lambda||_{l_p} 2^{(\frac{j}{2} - \frac{j}{p})}, \tag{8.15}$$

where $C_1 = (||\theta_\varphi||_\infty^{\frac{1}{p}} ||\varphi||_1^{\frac{1}{q}})^{-1}$, and $C_2 = ||\theta_\varphi||_\infty^{\frac{1}{q}} ||\varphi||_1^{\frac{1}{p}}$.

Proof First, observe that if $||\lambda||_{l_p} < \infty$, then $\sup_k |\lambda_k| < \infty$, and thus, under the Condition (θ) the series $\sum_k \lambda_k \varphi(x - k)$ is a.e. absolutely convergent.

$$|\sum_k \lambda_k \varphi(x - k)| \leq \sum_k |\lambda_k||\varphi(x - k)|^{\frac{1}{p}} |\varphi(x - k)|^{\frac{1}{q}}.$$

Using the Hölder inequality we get

$$\int |\sum_k \lambda_k \varphi(x - k)|^p dx \quad \leq \quad \int \sum_k |\lambda_k|^p |\varphi(x - k)| \{\sum_k |\varphi(x - k)|\}^{\frac{p}{q}} dx$$

$$\leq \quad ||\theta_\varphi||_\infty^{\frac{p}{q}} ||\lambda||_{l_p}^p \int |\varphi(x)| dx.$$

This yields (8.12) for $p < \infty$. For $p = \infty$ the proof is easier and left to the reader. Inequality (8.13) follows from (8.12) by renormalization. The right-hand side inequality in (8.14) coincides with (8.12). To prove the left-hand side inequality in (8.14) define $f(x) = \sum_k \lambda_k \varphi(x - k)$. Since φ is a scaling function, $\lambda_k = \int f(x)\varphi(x - k)dx$. Thus,

$$\sum_k |\lambda_k|^p \leq \sum_k \left(\int |f(x)||\varphi(x - k)|^{\frac{1}{p}} |\varphi(x - k)|^{\frac{1}{q}} dx \right)^p,$$

and by the Hölder inequality

$$\sum_k |\lambda_k|^p \leq \sum_k \int |f(x)|^p |\varphi(x-k)| dx \left(\int |\varphi(x-k)| dx \right)^{\frac{p}{q}}.$$

Hence,

$$\left(\sum_k |\lambda_k|^p \right)^{\frac{1}{p}} \leq \|\varphi\|_1^{\frac{1}{q}} \left(\int |f(x)|^p \sum_k |\varphi(x-k)| dx \right)^{\frac{1}{p}} \leq \|\varphi\|_1^{\frac{1}{q}} \|\theta_\varphi\|_\infty^{\frac{1}{p}} \|f\|_p.$$

This yields the proof for $p < \infty$. As above the case $p = \infty$ is left to the reader. Finally, (8.15) is a rescaled version of (8.14). □

If a scaling function satisfies Condition (θ), it is in some sense well concentrated. In this case the projection operator P_{V_0}, is given by a kernel operator with a periodic kernel.

PROPOSITION 8.4 *Let φ be a scaling function. If φ satisfies Condition (θ), then $P_{V_0}(f)(x) = Kf(x)$ for any $f \in L_2(I\!R)$, with*

$$K(x,y) = \sum_k \varphi(x-k)\overline{\varphi(y-k)}.$$

Proof Let $f \in L_2(I\!R)$. Then, by the Cauchy-Schwarz inequality,

$$\sum_k \left(\int |f(y)\overline{\varphi(y-k)}| dy \right) |\varphi(x-k)| \leq \sum_k \|f\|_2 \|\varphi\|_2 |\varphi(x-k)|$$

$$\leq \|f\|_2 \|\varphi\|_2 \theta_\varphi(x) < \infty.$$

So, by the Fubini theorem we have, for almost all x:

$$P_{V_0}(f)(x) = \sum_k \left(\int f(y)\overline{\varphi(y-k)} dy \right) \varphi(x-k)$$

$$= \int f(y) \sum_k \overline{\varphi(y-k)} \varphi(x-k) dy.$$

□

A very important fact here is that under Condition (θ), the projection operator P_{V_0} is given by a kernel $K(x,y)$ which acts also on other spaces than $L_2(I\!R)$, for instance, on all $L_p(I\!R)$, $1 \leq p \leq \infty$. If $f \in L_p(I\!R)$, clearly, by Hölder inequality we obtain that

$$\sum_k \left(\int |f(y)\overline{\varphi(y-k)}| dy \right) |\varphi(x-k)| \leq \|f\|_p \|\varphi\|_q \theta_\varphi(x),$$

where $\frac{1}{p} + \frac{1}{q} = 1$.

Proposition 8.4 justifies the following definition.

DEFINITION 8.7 (Orthogonal projection kernel). *Let φ be a scaling function satisfying Condition (θ). The kernel*

$$K(x,y) = \sum_k \varphi(x-k)\overline{\varphi(y-k)}$$

is called orthogonal projection kernel associated with φ.

REMARK 8.5 Obviously, the orthogonal projection kernel satisfies Condition P, i.e. it is periodic.

8.6   Moment condition for projection kernels

Here we specify the properties of φ necessary to obtain Condition $M(N)$ on the kernel

$$K(x,y) = \sum_k \varphi(x-k)\overline{\varphi(y-k)}.$$

First we formulate the properties of φ allowing to have various size conditions on K.

Condition S *(size condition) There exists a bounded non increasing function Φ such that*

$$\int \Phi(|u|)du < \infty,$$

and

$$|\varphi(u)| \leq \Phi(|u|) \quad (a.e.).$$

Condition $S(N)$ *Condition S holds and*

$$\int \Phi(|u|)|u|^N du < \infty.$$

LEMMA 8.5 *Condition (θ) follows from Condition S.*

Proof The function θ_φ is periodic, with period 1. Hence, Condition (θ) is satisfied if

$$\operatorname*{ess\,sup}_{x \in [0,1]} \theta_\varphi(x) < \infty. \tag{8.16}$$

But if $x \in [0,1]$, then $|x - k| \geq |k|/2$ for any $|k| \geq 2$. Hence, $\Phi(|x - k|) \leq \Phi(|k|/2)$, for any $|k| \geq 2$, $x \in [0,1]$. Using this, we get, under Condition S,

$$\begin{aligned}
\theta_\varphi(x) &= \sum_k |\varphi(x - k)| \leq \sum_k \Phi(|x - k|) \leq \Phi(|x|) + \Phi(|x + 1|) \\
&\quad + \Phi(|x - 1|) + \sum_{|k| \geq 2} \Phi(|k|/2) \leq 3\Phi(0) + \sum_k \Phi(|k|/2),
\end{aligned}$$

for almost all $x \in [0,1]$. Now, monotonicity of Φ yields

$$\sum_k \Phi(|k|/2) \leq \Phi(0) + \int_{-\infty}^{\infty} \Phi(|u|/2)\,du = C_\Phi < \infty. \tag{8.17}$$

Thus, (8.16) holds, which entails Condition (θ). $\qquad\square$

LEMMA 8.6 *If φ satisfies Condition S, then the kernel*

$$K(x, y) = \sum_k \varphi(x - k)\overline{\varphi(y - k)}$$

satisfies

$$|K(x, y)| \leq C_1 \Phi\left(\frac{|x - y|}{C_2}\right) \quad (a.e.),$$

where the positive constants C_1 and C_2 depend only on Φ.

Proof Using the monotonicity of Φ, we get, for any $n \in \mathbb{Z}$,

$$\begin{aligned}
\sum_k \Phi(|n - k|)\Phi(|k|) &\leq \sum_{|k| \leq |n|/2} \Phi(|n - k|)\Phi(|k|) + \sum_{|k| > |n|/2} \Phi(|n - k|)\Phi(|k|) \\
&\leq \Phi\left(\frac{|n|}{2}\right) \sum_{|k| \leq |n|/2} \Phi(|k|) + \Phi\left(\frac{|n|}{2}\right) \sum_k \Phi(|n - k|) \\
&\leq 2\Phi\left(\frac{|n|}{2}\right) \sum_k \Phi(|k|), \tag{8.18}
\end{aligned}$$

since $\sum_k \Phi(|n-k|) = \sum_k \Phi(|k|)$. As $\Phi(x/2)$ is also a monotone function, we get using (8.17) and (8.18),

$$\sum_k \Phi\left(\frac{|k|}{2}\right) \Phi\left(\frac{|n-k|}{2}\right) \leq 2C_\Phi \Phi\left(\frac{|n|}{4}\right). \qquad (8.19)$$

Any $x, y \in I\!\!R$ can be represented as

$$\begin{aligned} x &= k_0 + u \,, \; |u| \leq \tfrac{1}{2}, \\ y &= k_1 + v \,, \; |v| \leq \tfrac{1}{2}, \end{aligned}$$

where k_0 and k_1 are integers. Set $n = k_0 - k_1$. Then

$$\begin{aligned} |K(x,y)| &\leq \sum_k \Phi(|x-k|)\Phi(|y-k|) = \sum_k \Phi(|u-k|)\Phi(|v+n-k|) \\ &\leq \sum_k \Phi\left(\frac{|k|}{2}\right) \Phi\left(\frac{|n-k|}{2}\right) \leq 2C_\Phi \Phi\left(\frac{|n|}{4}\right), \end{aligned} \qquad (8.20)$$

where we used (8.19) and the inequalities $|u-k| \geq \frac{|k|}{2}$, $|v+n-k| \geq \frac{|n-k|}{2}$.

Let $\delta < \frac{1}{4}$ be such that $\Phi(\delta/2) > 0$. (If such δ does not exist, this means that $\Phi \equiv 0$, and the Lemma is trivial.) We have

$$\Phi\left(\frac{|n|}{4}\right) \leq \frac{\Phi(0)}{\Phi(\delta/2)} \Phi\left(\frac{\delta|x-y|}{2}\right). \qquad (8.21)$$

In fact, if $n \neq 0$, we have $2|n| \geq |n+u-v| = |x-y|$, and, by monotonicity of Φ,

$$\Phi\left(\frac{|n|}{4}\right) \leq \Phi(\delta|n|) \leq \Phi\left(\frac{\delta|x-y|}{2}\right).$$

If $n = 0$, then $|x-y| = |u-v| \leq 1$, and

$$\Phi\left(\frac{|n|}{4}\right) = \Phi(0) \leq \frac{\Phi(0)}{\Phi(\delta/2)} \Phi\left(\frac{\delta|x-y|}{2}\right).$$

Combining (8.20) and (8.21), we obtain the Lemma.   □

Using Lemma 8.6, it is easy to see that, Condition $S(N)$ being satisfied, the Condition $H(N)$ holds as well, and the following quantities are well-defined

$$m_n = \int \varphi(x)x^n dx,$$
$$\mu_n(t) = \int K(t,s)(s-t)^n ds,$$
$$C_n(t) = \sum_k \varphi(t-k)(t-k)^n, \quad n=0,1,\ldots,N.$$

PROPOSITION 8.5 *Let, for some $N \geq 0$, φ satisfy Condition $S(N)$ and $\int \varphi(x)dx \neq 0$. Then K, associated with φ, satisfies Conditions P and $H(N)$, and we have the following.*

(i) $\mu_n(t) = \sum_{j=0}^{n}(-1)^{n-j}\binom{n}{j}\overline{m_j}C_{n-j}(t), \quad n=0,1,\ldots,N.$

(ii) *The following three relations are equivalent:*

$$C_n(t) = C_n \quad (a.e.), \quad n=0,1,\ldots,N, \tag{8.22}$$
$$\mu_n(t) = \mu_n \quad (a.e.), \quad n=0,1,\ldots,N, \tag{8.23}$$
$$\hat{\varphi}(\xi+2k\pi) = o(|\xi|^N), \text{ as } \xi \to 0, \forall k \neq 0, \tag{8.24}$$

where C_n and μ_n are some constants. Each of these relations implies that
$$C_n = m_n, \quad n=0,1,\ldots,N, \tag{8.25}$$
and
$$\mu_n = \int (-t)^n (\varphi * \check{\varphi})(t)dt, \quad n=0,1,\ldots,N, \tag{8.26}$$
where $\check{\varphi}(t) = \overline{\varphi}(-t)$.

(iii) *The kernel K satisfies Condition $M(N)$ if and only if (8.24) holds and $|\hat{\varphi}(\xi)|^2 = 1 + o(|\xi|^N)$, as $\xi \to 0$.*

(iv) *In particular, if φ satisfies the condition S, then we have:*

$$K \text{ satisfies } M(0) \Leftrightarrow \hat{\varphi}(2k\pi) = \delta_{0k}, \forall k \in \mathbb{Z}.$$

Proof

(i) By the binomial formula

$$
\begin{aligned}
\mu_n(t) &= \int K(t,s)(s-t)^n ds \\
&= \sum_k \int \varphi(t-k)\overline{\varphi(s-k)}(s-k+k-t)^n ds \\
&= \sum_{j=0}^{n} (-1)^{n-j} \binom{n}{j} \overline{m}_j C_{n-j}(t).
\end{aligned}
$$

(ii) It follows from (i) that (8.22) $\Rightarrow$ (8.23). The inverse implication is proved by induction. In fact, if (8.23) holds, we have $\mu_0 = \overline{m}_0 C_0(t) = (\int \varphi(x)dx) C_0(t)$. Thus, $C_0(t) = C_0 = \mu_0/\overline{m}_0, \forall t$. Next, assume that (8.23) entails (8.22) for $n = 0, 1, \ldots, N-1$, and observe that it entails (8.22) also for $n = N$, in view of (i).

It remains to show the equivalence of (8.22) and (8.24). By the property (4.9) of the Fourier transforms (see Chapter 4), we have

$$
\hat{\varphi}^{(n)}(\xi) = \int \varphi(t)(-it)^n e^{-i\xi t} dt.
$$

In particular,

$$
\hat{\varphi}^{(n)}(2k\pi) = \int \varphi(t)(-it)^n e^{-i2k\pi t} dt. \tag{8.27}
$$

and by (4.10) and the Poisson summation formula (4.13) of Chapter 4, with $T = 1$,

$$
\begin{aligned}
\hat{\varphi}^{(n)}(2k\pi) &= \int_0^1 \sum_{m=-\infty}^{+\infty} \varphi(t-m)\{-i(t-m)\}^n e^{-i2k\pi t} dt \\
&= (-i)^n \int_0^1 C_n(t) e^{-i2k\pi t} dt. \tag{8.28}
\end{aligned}
$$

Note that (8.24) is equivalent to

$$
\hat{\varphi}^{(n)}(2k\pi) = 0, \quad n = 0, 1, \ldots, N, \ k \neq 0. \tag{8.29}
$$

But, in view of (8.28), the condition (8.29) holds if and only if $C_n(t)$ is constant for all $t \in [0,1]$ (note that by (8.28) the Fourier coefficients of $C_n(t)$ on $[0,1]$ are proportional to $\hat{\varphi}^{(n)}(2k\pi)$). Thus, (8.22) is equivalent to (8.24).

To prove that $(8.23) \Rightarrow (8.25)$ we apply (8.28) with $k = 0$. We get

$$\hat{\varphi}^{(n)}(0) = (-i)^n \int_0^1 C_n(t)dt \equiv (-i)^n C_n.$$

On the other hand, $\hat{\varphi}^{(n)}(0) = (-i)^n m_n$ by (8.27). Thus, (8.25) follows. The proof of (8.26) is given by the next calculations.

$$
\begin{aligned}
\mu_n &= \sum_{j=0}^{n} (-1)^{n-j} \binom{n}{j} \overline{m}_j m_{n-j} \\
&= \sum_{j=0}^{n} (-1)^{n-j} \binom{n}{j} \int v^j \overline{\varphi(v)} dv \int u^{n-j} \varphi(u) du \\
&= \int \int \sum_{j=0}^{n} \binom{n}{j} v^j (-u)^{n-j} \overline{\varphi(v)} \varphi(u) du dv \\
&= \int \int (v-u)^n \overline{\varphi(v)} \varphi(u) du dv \\
&= \int (-t)^n (\varphi * \tilde{\varphi})(t) dt.
\end{aligned}
\tag{8.30}
$$

(iii) The condition (8.3) may be rewritten as

$$\mu_0(t) \equiv 1, \quad \mu_n(t) \equiv 0, \quad n = 1, \ldots, N, \tag{8.31}$$

which is a special case of (8.23). But $(8.23) \Rightarrow (8.26)$. Using (8.26), we rewrite (8.31) as

$$
\begin{aligned}
\mathcal{F}[\varphi * \tilde{\varphi}](0) &= \int (\varphi * \tilde{\varphi})(t) dt = 1, \\
\mathcal{F}^{(n)}[\varphi * \tilde{\varphi}](0) &= \int (-it)^n (\varphi * \tilde{\varphi})(t) dt = 0, \quad n = 1, \ldots, N, \tag{8.32}
\end{aligned}
$$

where $\mathcal{F}^{(n)}$ is the nth derivative of the Fourier transform $\mathcal{F}$. By the property (4.8) of Fourier transforms (see Chapter 4), $\mathcal{F}[\varphi * \tilde{\varphi}](\xi) = |\hat{\varphi}(\xi)|^2$. Therefore, (8.32) is equivalent to $|\hat{\varphi}(\xi)|^2 = 1 + o(|\xi|^N)$ as $\xi \to 0$. This implies that (8.3) holds if and only if (8.23) is true and $|\hat{\varphi}(\xi)|^2 = 1 + o(|\xi|^N)$ as $\xi \to 0$. To finish the proof note that $(8.23) \Leftrightarrow (8.24)$ by (ii) of this proposition.

(iv) Is obvious.

We finish this section with the following remark related to the condition $M(N)$ in the simplest case of a convolution kernel.

REMARK 8.6 If $K(x,y) = K^*(x - y)$ is a convolution kernel and $K^* \in L_1(\mathbb{R})$, then

$$K \text{ satisfies Condition } M(N) \Leftrightarrow$$

$$\int |x|^N |K^*(x)| dx < \infty \text{ and } \hat{K}^*(\xi) = 1 + o(|\xi|^N), \text{ as } \xi \to 0.$$

8.7   Moment condition in the wavelet case

Proposition 8.5 explains how to guarantee the Condition $M(N)$ for an orthogonal projection kernel $K(x,y) = \sum_k \varphi(x-k)\overline{\varphi(y-k)}$. Let us now investigate what can be improved, if φ is a father wavelet that generates a MRA. The definition of MRA was given in Chapter 3. It contained the following three conditions on φ:

- $\{\varphi(x - k), \ k \in \mathbb{Z}\}$ is an ONS,

- the spaces V_j are nested: $V_j \subset V_{j+1}$,

- $\bigcup_{j \geq 0} V_j$ is dense in $L_2(\mathbb{R})$, where V_j is the linear subspace of $L_2(\mathbb{R})$ spanned by $\{2^{j/2}\varphi(2^j x - k), \ k \in \mathbb{Z}\}$.

Here it will be sufficient to impose only the first two of these conditions, since we work in this section under the strong Condition $S(N)$. The fact that $\bigcup_{j \geq 0} V_j$ is dense in $L_2(\mathbb{R})$ will follow as a consequence (see Corollary 8.1 below).

In view of Lemma 5.1, the fact that $\{\varphi(x - k), \ k \in \mathbb{Z}\}$ is an ONS may be expressed by the relation

$$\sum_k |\hat{\varphi}(\xi + 2k\pi)|^2 = 1 \qquad (a.e.), \tag{8.33}$$

and, by Proposition 5.1, the spaces V_j are nested if and only if

$$\hat{\varphi}(\xi) = \hat{\varphi}\left(\frac{\xi}{2}\right) m_0\left(\frac{\xi}{2}\right) \qquad (a.e.), \tag{8.34}$$

where $m_0(\xi)$ is a 2π-periodic function, $m_0 \in L_2(0, 2\pi)$.

REMARK 8.7 If the scaling function φ satisfies Condition $S(N)$, for some $N \geq 0$, then the orthogonal projection operator P_{V_j} on V_j is given by the kernel

$$K_j(x,y) = 2^j K(2^j x, 2^j y) = \sum_k 2^{\frac{1}{2}} \varphi(2^j x - k) 2^{\frac{1}{2}} \overline{\varphi(2^j y - k)}.$$

In fact, Condition $S(N)$ implies Condition (θ) (Lemma 8.5), and one can apply Proposition 8.4 with obvious rescaling of φ.

Let us recall that if P and Q are two operators given by two kernels, $K(x,y)$ and $F(x,y)$, then the composed operator $P \circ Q$ is given by the composed kernel $K \circ F(x,y) = \int K(x,z)F(z,y)dz$. Since the spaces V_j are nested, we have $P_{V_j} \circ P_{V_0} = P_{V_0}, j = 1, 2, \ldots$.

THEOREM 8.2 *Let φ be a scaling function satisfying (8.33), (8.34) and Condition $S(N)$. If $\varphi \in \check{W}_q^N(\mathbb{R})$ for some integer $N \geq 0$ and some $1 \leq q \leq \infty$, then the kernel $K(x,y) = \sum_k \varphi(x-k)\overline{\varphi(y-k)}$ satisfies the moment condition $M(N)$.*

Proof Note that $K_j \varphi = \varphi$ for $j = 1, 2, \ldots$. In fact, by the property of projection operators mentioned above, $P_{V_j}(\varphi) = P_{V_j} \circ P_{V_0}(\varphi) = P_{V_0}(\varphi) = \varphi$, since $\varphi \in V_0$. Also, φ is not a constant, since $\varphi \in L_2(\mathbb{R})$. Thus, the assumptions of Theorem 8.1 (iii) are fulfilled for $f = \varphi$, $h = 2^{-j}$, and K satisfies Condition $M(N)$. □

This theorem gives a sufficient condition. Let us now derive a necessary and sufficient condition for the Condition $M(N)$. We shall show that, if K is the projection operator on the space V_0 of a multiresolution analysis then it is possible to improve Proposition 8.5.

First, we state properties of multiresolution analysis under the Condition (θ) on the father wavelet φ. For this recall some notation from Chapters 3 and 5. Let

$$m_1(\xi) = \overline{m_0(\xi + \pi)}e^{-i\xi}, \tag{8.35}$$

$$\hat{\psi}(\xi) = m_1\left(\frac{\xi}{2}\right)\hat{\varphi}\left(\frac{\xi}{2}\right), \tag{8.36}$$

and let the mother wavelet ψ be the inverse Fourier transform of $\hat{\psi}$. Let W_0 be the orthogonal complement of V_0 in V_1, i.e. $V_1 = V_0 \oplus W_0$.

PROPOSITION 8.6 *Let φ be a scaling function satisfying (8.33), (8.34) and the Condition (θ). Then*

(i) **For all** ξ

$$\sum_k |\hat{\varphi}(\xi + 2k\pi)|^2 = 1.$$

(ii) *The function m_0 is a 2π-periodic continuous function with absolutely convergent Fourier series.*

(iii) $m_0(0) = 1, |\hat{\varphi}(0)| = 1, \hat{\varphi}(2k\pi) = 0, \forall k \neq 0$.

(iv) $\{\psi(x - k),\ k \in \mathbb{Z}\}$ *is an ONB in W_0.*

(v) *The mother wavelet ψ satisfies the Condition (θ). If, moreover, $\int |x|^N |\varphi(x)| dx < \infty$, then $\int |x|^N |\psi(x)| dx < \infty$.*

(vi) *Let $D(x, y) = K_1(x, y) - K(x, y)$. Then D is the kernel of the orthogonal projection operator on W_0, and we have*

$$D(x, y) = \sum_k \psi(x - k)\overline{\psi(y - k)}.$$

Proof

(i) Fix ξ and define the function

$$g_\xi(x) = \sum_{n=-\infty}^{\infty} \varphi(x + n) \exp\{-i\xi(x + n)\}.$$

The function $g_\xi(x)$ is bounded, in view of Condition (θ), and it is periodic, with period 1. By the Poisson summation formula ((4.13) of Chapter 4, with $T = 1$) the Fourier coefficients of $g_\xi(x)$ are $\hat{\varphi}(\xi + 2k\pi), k \in \mathbb{Z}$.

To prove (i) we proceed now as in Lemarié (1991). By Parseval's formula

$$\sum_k |\hat{\varphi}(\xi + 2k\pi)|^2 = \int_0^1 |g_\xi(x)|^2 dx, \forall \xi \in \mathbb{R}.$$

The RHS of this equation is a continuous function of ξ since g_ξ is a bounded continuous function. Hence, $\sum_k |\hat{\varphi}(\xi + 2k\pi)|^2$ is a continuous function of ξ, which, together with (8.33), proves (i).

(ii) Using the argument after formula (5.3) of Chapter 5, we find that the function $m_0(\xi)$ in (8.34) may be written as

$$m_0(\xi) = \sum_k a_k e^{-ik\xi}$$

with $a_k = \int \varphi(x)\overline{\varphi(2x - k)}dx$, where

$$\sum_k |a_k| \le \int \sum_k |\varphi(x)||\varphi(2x - k)|dx \le \|\theta_\varphi\|_\infty \|\varphi\|_1 < \infty.$$

(iii) Lemma 5.2 of Chapter 5 yields that, under (8.33) and (8.34), $|m_0(\xi)|^2 + |m_0(\xi + \pi)|^2 = 1$ (*a.e.*). This equality is true everywhere, since by (ii) m_0 is continuous. Thus, $|m_0(0)| \le 1$. Let us show that $|m_0(0)| = 1$. In fact, if $|m_0(0)| < 1$, then $|m_0(\xi)| < \eta < 1$, for ξ small enough, and, for any $\xi \in \mathbb{R}$,

$$\hat{\varphi}(\xi) = \hat{\varphi}(\frac{\xi}{2})m_0(\frac{\xi}{2}) = \hat{\varphi}(\frac{\xi}{2^{q+1}})m_0(\frac{\xi}{2^{q+1}})\dots m_0(\frac{\xi}{2}) \to 0 \text{ as } q \to \infty.$$

Thus, $\hat{\varphi}(\xi) = 0$, $\forall \xi \in \mathbb{R}$, which is impossible. Hence, $|m_0(0)| = 1$. Also, $|m_0(2k\pi)|^2 = 1$, $k \in \mathbb{Z}$, by periodicity of m_0. Using this and applying (8.34), we obtain

$$\begin{aligned}|\hat{\varphi}(2^j 2k\pi)| &= |\hat{\varphi}(2^{j-1}2k\pi)||m_0(2^{j-1}2k\pi)| \\ &= |\hat{\varphi}(2^{j-1}2k\pi)|, \ k \in \mathbb{Z}, \ j = 1, 2, \dots.\end{aligned}$$

Hence, for any $k \in \mathbb{Z}$,

$$|\hat{\varphi}(2^j 2k\pi)| = |\hat{\varphi}(2k\pi)|, \quad j = 1, 2, \dots. \tag{8.37}$$

Fix $k \ne 0$. Take limits of both sides of (8.37), as $j \to \infty$, and note that by Riemann-Lebesgue Lemma we have $\hat{\varphi}(\xi) \to 0$, as $|\xi| \to \infty$. We obtain $\hat{\varphi}(2k\pi) = 0$, $k \ne 0$. This, and (8.33) imply that $|\hat{\varphi}(0)| = 1$. Now, (8.34) entails that $m_0(0) = 1$.

(iv) See Lemma 5.3 and Remark 5.2 of Chapter 5.

(v) The mother wavelet $\psi(x)$ may be written as (cf. (5.13) and the relation $h_k = \sqrt{2}a_k$, see the definition of h_k after (5.3) in Chapter 5):

$$\psi(x) = \sqrt{2}\sum_k (-1)^{k+1}\bar{h}_{1-k}\varphi(2x - k)$$
$$= 2\sum_k (-1)^k \bar{a}_k \varphi(2x - 1 + k).$$

Thus, the Condition (θ) on the function ψ follows from the inequalities

$$\sum_l |\psi(x - l)| \leq 2\sum_k \sum_l |a_k||\varphi(2x - 2l - 1 + k)|$$
$$\leq 2\sum_k |a_k| \sum_l |\varphi(2x - 2l - 1 + k)|$$
$$\leq 2\|\theta_\varphi\|_\infty \sum_k |a_k|.$$

Next, suppose that $\int |x|^N |\varphi(x)|dx < \infty$. Then

$$\int |\psi(x)||x|^N dx \leq \sum_k 2|a_k| \int |\varphi(2x - 1 + k)||x|^N dx$$
$$\leq C\sum_k |a_k| \int |\varphi(x)|(|x|^N + |k|^N)dx,$$

where $C > 0$ is a constant. It remains to prove that $\sum_k |a_k||k|^N < \infty$. We have

$$\sum_k |a_k||k|^N \leq \int \sum_k |\varphi(x)||\varphi(2x - k)||k|^N dx$$
$$\leq \tilde{C}\sum_k \int |\varphi(x)||\varphi(2x - k)|(|2x - k|^N + |x|^N)dx$$
$$\leq C'\|\theta_\varphi\|_\infty \int |x|^N |\varphi(x)|dx < \infty,$$

where $\tilde{C}$ and C' are positive constants.

(vi) The system $\{\psi(x - k),\ k \in \mathbb{Z}\}$ is an ONB of W_0 in view of (iv). The function ψ satisfies Condition (θ) in view of (v). Hence, we can apply Proposition 8.4, with W_0 instead of V_0 and ψ instead of φ.

□

COROLLARY 8.1 *Let φ be a scaling function, satisfying (8.33), (8.34) and the Condition S. Then*

(i) The associated orthogonal projection kernel

$$K(x,y) = \sum_k \varphi(x-k)\overline{\varphi(y-k)}$$

satisfies the Condition $M(0)$, i.e. $\int K(x,y)dy = 1$.

(ii) $\bigcup_{j \geq 0} V_j$ is dense in $L_2(\mathbb{R})$.

Proof

(i) By Proposition 8.5 (iii) it suffices to verify that $\hat{\varphi}(\xi + 2k\pi) = o(1)$, as $\xi \to 0$, $\forall k \neq 0$, and $|\hat{\varphi}(\xi)|^2 = 1 + o(1)$, as $\xi \to 0$. But these relations follow from Proposition 8.6 (iii) and from the obvious fact that $\hat{\varphi}(\cdot)$ is a continuous function under the Condition S.

(ii) It suffices to show that $||P_{V_j}(f) - f||_2 \to 0$, for any $f \in L_2(\mathbb{R})$, as $j \to \infty$. This follows from Theorem 8.1 (i) applied for $N = 0$, $p = 2$, $h = 2^{-j}$. In fact, the assumptions of Theorem 8.1 (i) are satisfied in view of Remark 8.7, the point (i) of the present Corollary and of the fact that

$$L_2(\mathbb{R}) = \tilde{W}_2^0(\mathbb{R}).$$

□

Here is now the main theorem of this section, which is a refinement of Proposition 8.5 in the context of multiresolution analysis.

THEOREM 8.3 *Let φ be a scaling function, satisfying (8.33), (8.34) and the Condition $S(N)$ for some integer $N \geq 0$. Let $K(x,y)$ be the associated orthogonal projection kernel, and let ψ be the associated mother wavelet defined by (8.35) and (8.36).*

The following properties are equivalent:

(i) $|m_0(\xi)|^2 = 1 + o(|\xi|^{2N})$, as $\xi \to 0$,

(ii) $\int x^n \psi(x)dx = 0, \quad n = 0, 1, \dots, N$,

(iii) $\hat{\varphi}(\xi + 2k\pi) = o(|\xi|^N)$, *as* $\xi \to 0$, $\forall k \neq 0$,

(iv) $K(x, y)$ *satisfies the Condition* $M(N)$.

If, moreover, the function $|\hat{\varphi}(\xi)|^2$ is $2N$ times continuously differentiable at $\xi = 0$, then the properties (i) - (iv) are equivalent to

$$|\hat{\varphi}(\xi)|^2 = 1 + o(|\xi|^{2N}), \quad as \ \xi \to 0. \tag{8.38}$$

REMARK 8.8 The property (i) is equivalent to

$$m_0(\xi + \pi) = o(|\xi|^N), \quad as \ \xi \to 0. \tag{8.39}$$

and to

$$m_1(\xi) = o(|\xi|^N), \quad as \ \xi \to 0. \tag{8.40}$$

In fact, by Lemma 5.2 of Chapter 5,

$$|m_0(\xi)|^2 + |m_0(\xi + \pi)|^2 = 1 \quad (a.e.). \tag{8.41}$$

Moreover, (8.41) holds for all ξ (not only a.e.), since in view of Proposition 8.6 (i), we can skip (a.e.) in (8.33). This implies that (i) of Theorem 8.3 and (8.39) are equivalent. The equivalence of (8.39) and (8.40) follows from the definition of $m_1(\xi)$ (see (8.35)).

REMARK 8.9 The function $|\hat{\varphi}(\xi)|^2$ is $2N$ times continuously differentiable, if e.g. $\int |t|^{2N} |\varphi(t)| dt < \infty$. This is always the case for compactly supported φ.

Proof of Theorem 8.3

(i)$\Leftrightarrow$ (ii) Note that (ii) is equivalent to the relation $\hat{\psi}(\xi) = o(|\xi|^N), \xi \to 0$, by the property of derivatives of Fourier transforms (Chapter 4, formula (4.9)). Now, $\hat{\psi}(\xi) = m_1(\frac{\xi}{2})\hat{\varphi}(\frac{\xi}{2}), \hat{\varphi}(0) \neq 0$ by Proposition 8.6 (iii), and $\hat{\varphi}(\xi)$ is continuous. Hence, $\hat{\psi}(\xi) = o(|\xi|^N), \ \xi \to 0, \Leftrightarrow$ (8.40) holds. Finally, (8.40) $\Leftrightarrow$ (i) by Remark 8.8.

(i)$\Rightarrow$ (iii) Using Remark 8.8, we can replace (i) by (8.39). Now, any $k \in \mathbb{Z}$, $k \neq 0$, may be represented as $k = 2^q k'$, where k' is odd, and $q \geq 0$ is an integer. Hence,

$$
\begin{aligned}
\hat{\varphi}(\xi + 2k\pi) &= \hat{\varphi}(\frac{\xi}{2} + k\pi)m_0\left(\frac{\xi}{2} + k\pi\right) \\
&= \hat{\varphi}(\frac{\xi}{2^{q+1}} + k'\pi)m_0\left(\frac{\xi}{2^{q+1}} + k'\pi\right) \ldots m_0(\frac{\xi}{2} + k\pi).
\end{aligned}
$$

As m_0 is 2π-periodic and (8.39) holds, we obtain

$$
m_0(\frac{\xi}{2^{q+1}} + k'\pi) = m_0(\frac{\xi}{2^{q+1}} + \pi) = o(|\xi|^N), \quad \text{as } \xi \to 0.
$$

Using this and the fact that $\hat{\varphi}$ and m_0 are uniformly bounded ($|m_0(\xi)| \leq 1$, by (8.41)), we get (iii).

(iii)$\Rightarrow$ (i) Proposition 8.6 (i) guarantees the existence of such k_0 that

$$
\hat{\varphi}(\pi + 2k_0\pi) \neq 0. \tag{8.42}
$$

Let $k_0' = 2k_0 + 1$. Then, for every ξ,

$$
\hat{\varphi}(\xi + 2k_0'\pi) = m_0\left(\frac{\xi}{2} + k_0'\pi\right)\hat{\varphi}(\frac{\xi}{2} + k_0'\pi) = m_0\left(\frac{\xi}{2} + \pi\right)\hat{\varphi}(\frac{\xi}{2} + \pi + 2k_0\pi),
$$

where we used the fact that m_0 is 2π-periodic. Letting in this relation $\xi \to 0$ and using (iii), the continuity of $\hat{\varphi}$ and (8.42) we get $m_0(\xi + \pi) = o(|\xi|^N)$, which, in view of Remark 8.8, is equivalent to (i).

(ii)$\Leftrightarrow$ (iv) By Proposition 8.5 (iii) it suffices to show that (iii) implies

$$
|\hat{\varphi}(\xi)|^2 = 1 + o(|\xi|^N), \quad \text{as } \xi \to 0. \tag{8.43}
$$

To show this, note that (iii) $\Rightarrow$ (i), and thus

$$
\begin{aligned}
|\hat{\varphi}(\xi)|^2 &= |\hat{\varphi}(\frac{\xi}{2})|^2 |m_0(\frac{\xi}{2})|^2 \\
&= |\hat{\varphi}(\frac{\xi}{2})|^2 (1 + o(|\xi|^{2N})). \tag{8.44}
\end{aligned}
$$

as $\xi \to 0$. Next, note that $|\hat{\varphi}(\xi)|^2$ is N times continuously differentiable at $\xi = 0$. In fact, $|\hat{\varphi}(\xi)|^2$ is the Fourier transform of the function $\varphi * \tilde{\varphi}$ (see (4.8) of Chapter 4), and derivative of order $n \leq N$ of $|\hat{\varphi}(\xi)|^2$ at $\xi = 0$ is

$$\frac{d^n|\hat{\varphi}(\xi)|^2}{d\xi^n}\bigg|_{\xi=0} = \int(-it)^n(\varphi * \tilde{\varphi})(t)dt$$

$$= i^n\mu_n,$$

where we used the property of Fourier transforms (4.9) of Chapter 4, and (8.26). Also, $|\hat{\varphi}(0)|^2 = 1$ by Proposition 8.6 (iii). Hence, there exist numbers $b_1, \ldots, b_N$ such that the Taylor expansion holds:

$$|\hat{\varphi}(\xi)|^2 = 1 + \sum_{k=1}^{N} b_k\xi^k + o(|\xi|^N), \tag{8.45}$$

as $\xi \to 0$. Combining (8.44) and (8.45) we get

$$1 + \sum_{k=1}^{N} b_k\xi^k + o(|\xi|^N) = (1 + o(|\xi|^{2N}))\left(1 + \sum_{k=1}^{N} b_k(\frac{\xi}{2})^k + o(|\xi|^N)\right),$$

which implies $b_1 = \ldots = b_N = 0$, and, consequently, (8.43).

(iii)$\Leftrightarrow$ (8.38) Since $|\hat{\varphi}(\xi)|^2$ is $2N$ times differentiable the proof of (iii)$\Leftrightarrow$ (8.38) is similar to the proof of (iii) $\Leftrightarrow$ (iv), and is therefore omitted.

(8.38)$\Rightarrow$ (i) is obvious.

$$\square$$

REMARK 8.10 Comparison of Proposition 8.5 and Theorem 8.3.
If φ is a general scaling function, as in Proposition 8.5, then the two characteristic properties, guaranteeing Condition $M(N)$, i.e.

- $\hat{\varphi}(\xi + 2k\pi) = o(|\xi|^N)$, as $\xi \to 0, \forall k \neq 0$, k integer,

  and

- $|\hat{\varphi}(\xi)|^2 = 1 + o(|\xi|^N)$, as $\xi \to 0$, are independent. But if φ is a scaling function of a multiresolution analysis (in other words, φ is a father wavelet), then the first property implies the second. This is the case considered in Theorem 8.3.

The following corollary summarizes the results of this chapter. It presents explicitly the approximation properties of wavelet expansions on the Sobolev spaces.

COROLLARY 8.2 *Let φ be a scaling function satisfying (8.33), (8.34) and the Condition $S(N+1)$, for some integer $N \geq 0$. Let, in addition, at least one of the following four assumptions hold:*

(W1) $\varphi \in \tilde{W}_q^N(\mathbb{R})$ for some $1 \leq q \leq \infty$,

(W2) $|m_0(\xi)|^2 = 1 + o(|\xi|^{2N})$, as $\xi \to 0$,

(W3) $\int x^n \psi(x)dx = 0$, $n = 0, 1, \ldots, N$, where ψ is the mother wavelet associated to φ,

(W4) $\hat{\varphi}(\xi + 2k\pi) = o(|\xi|^N)$, as $\xi \to 0, \forall k \neq 0$.

Then, if f belongs to the Sobolev space $W_p^{N+1}(\mathbb{R})$, we have

$$\|K_j f - f\|_p = O\left(2^{-j(N+1)}\right), \quad as \ j \to \infty, \tag{8.46}$$

for any $p \in [1, \infty]$, where K_j is the wavelet projection kernel on V_j,

$$K_j(x, y) = \sum_k 2^j \varphi(2^j x - k)\overline{\varphi(2^j y - k)}.$$

Proof By Theorems 8.2 and 8.3, the Condition $M(N)$ is satisfied for $K(x, y)$, the orthogonal projection kernel associated with φ. Moreover, by Lemma 8.6 Condition $S(N+1)$ implies Condition $H(N+1)$. It remains to apply Theorem 8.1 (ii) with $h = 2^{-j}$. $\quad\square$

In view of this corollary, the simplest way to obtain the approximation property (8.46) is to use a compactly supported father wavelet φ that is smooth enough. This ensures both Condition $S(N+1)$ and (W1). However, the condition (W1) is not always the easiest to check, and the conditions (W2) to (W4) (all these three conditions, as shown in Theorem 8.3, are equivalent) may be more convenient. Note that (W2) to (W4) are necessary and sufficient conditions, while (W1) is a more restrictive assumption, as the following example shows.

EXAMPLE 8.1 Consider the Daubechies $D2(N+1)$ father wavelet $\varphi = \varphi_{D2(N+1)}$. For this wavelet we have (see (7.5) of Chapter 7)

$$
\begin{aligned}
|m_0(\xi)|^2 &= c_N \int_{\xi}^{\pi} \sin^{2N+1} x\, dx \\
&= 1 + O(|\xi|^{2N+2}), \text{ as } \xi \to 0,
\end{aligned}
$$

which yields (W2). Also, we know that $\varphi_{D2(N+1)}$ is bounded and compactly supported. By Theorem 8.3, the corresponding projection kernel $K(x, y)$ satisfies Condition $M(N)$, and by Corollary 8.2 we have the approximation property (8.46). But (W1) is not satisfied: there is no $q \geq 1$ such that $\varphi_{D2(N+1)} \in W_q^N$. This shows that Theorem 8.3 is stronger than Theorem 8.2.

Chapter 9

Wavelets and Besov Spaces

9.1 Introduction

This chapter is devoted to approximation theorems in Besov spaces. The advantage of Besov spaces as compared to the Sobolev spaces is that they are much more general tool in describing the smoothness properties of functions. We show that Besov spaces admit a characterization in terms of wavelet coefficients, which is not the case for Sobolev spaces. Thus the Besov spaces are intrinsically connected to the analysis of curves via wavelet techniques. The results of Chapter 8 are substantially used throughout. General references about Besov spaces are Nikol'skii (1975), Peetre (1975), Besov, Il'in & Nikol'skii (1978), Bergh & Löfström (1976), Triebel (1992), DeVore & Lorentz (1993).

9.2 Besov spaces

In this section we give the definition of the Besov spaces. We start by introducing the moduli of continuity of first and second order, and by discussing some of their properties.

DEFINITION 9.1 (Moduli of continuity.) *Let f be a function in $L_p(\mathbb{R}), 1 \le p \le \infty$. Let $\tau_h f(x) = f(x - h)$, $\Delta_h f = \tau_h f - f$. We define also $\Delta_h^2 f = \Delta_h \Delta_h f$. For $t \ge 0$ the moduli of continuity are defined by*

$$\omega_p^1(f, t) = \sup_{|h| \le t} \|\Delta_h f\|_p, \quad \omega_p^2(f, t) = \sup_{|h| \le t} \|\Delta_h^2 f\|_p.$$

The following lemma is well known, see DeVore & Lorentz (1993, Chapter 2).

LEMMA 9.1 *For f in $L_p(\mathbb{R})$, we have:*

(i) $\omega_p^1(f,t)$, *and* $\omega_p^2(f,t)$ *are non-decreasing functions of t and,* $\omega_p^2(f,t) \leq 2\omega_p^1(f,t) \leq 4\|f\|_p$,

(ii) $\omega_p^1(f,t) \leq \sum_{j=0}^{\infty} 2^{-(j+1)}\omega_p^2(f,2^jt) \leq t\int_t^{\infty} \frac{\omega_p^2(f,s)}{s^2}ds$ *(the* **Marchaud inequality**),

(iii) $\omega_p^1(f,ts) \leq (s+1)\omega_p^1(f,t)$, *for any* $s \geq 0$, $t \geq 0$,

(iv) $\omega_p^2(f,ts) \leq (s+1)^2\omega_p^2(f,t)$, *for any* $s \geq 0$, $t \geq 0$,

(v) $\omega_p^1(f,t) \leq t\|f'\|_p$, *if* $f \in W_p^1(\mathbb{R})$,

(vi) $\omega_p^2(f,t) \leq t^2\|f''\|_p$, *if* $f \in W_p^2(\mathbb{R})$.

Proof

(i) This is an obvious consequence of the definition.

(ii) We observe that $2\Delta_h = \Delta_{2h} - \Delta_h^2$. This implies: $\omega_p^1(f,t) \leq \frac{1}{2}(\omega_p^2(f,t) + \omega_p^1(f,2t))$, and thus

$$\omega_p^1(f,t) \leq \sum_{j=0}^{k} 2^{-(j+1)}\omega_p^2(f,2^jt) + 2^{-(k+1)}\omega_p^1(f,2^{(k+1)}t).$$

This yields the first inequality in (ii) if we let $k \to \infty$. The second inequality follows from the comparison of the series and the Riemann integral (note that $\omega_p^2(f,s)$ is non-decreasing in s and $\frac{1}{s^2}$ is decreasing).

(iii) Note that $\omega_p^1(f,t)$ is a subadditive function of t, so that $\omega_p^1(f,nt) \leq n\omega_p^1(f,t)$ for any integer n.

(iv) We have $\Delta_{nh}f(x) = \sum_{k=0}^{n-1}\Delta_h f(x - kh)$. Then

$$\Delta_{nh}^2 f(x) = \sum_{k'=0}^{n-1}\sum_{k=0}^{n-1}\Delta_h^2 f(x - kh - k'h).$$

Thus, $\omega_p^2(f,nt) \leq n^2\omega_p^2(f,t)$ for any integer n.

(v) If $f \in W_p^1$, we have $\Delta_h f(x) = f(x-h) - f(x) = -h \int_0^1 f'(x-sh)ds$, and $\|\Delta_h f\|_p \le |h| \|f'\|_p$.

(vi) Let $f \in W_p^2$. Then

$$f(x-2h) - f(x-h) = -f'(x-h)h + \frac{h^2}{2} \int_0^1 f''(x-h-sh)ds.$$

Quite similarly,

$$f(x) - f(x-h) = f'(x-h)h + \frac{h^2}{2} \int_0^1 f''(x-h+sh)ds.$$

Thus,

$$\begin{aligned} \Delta_h^2 f(x) &= f(x-2h) - 2f(x-h) + f(x) \\ &= \frac{h^2}{2} \int_0^1 \{f''(x-h+sh) + f''(x-h-sh)\}ds. \end{aligned}$$

Therefore,

$$\begin{aligned} \|\Delta_h^2 f\|_p &\le h^2 \left[\int_{-\infty}^\infty \int_0^1 \left| \frac{f''(x-h+sh) + f''(x-h-sh)}{2} \right|^p dsdx \right]^{1/p} \\ &\le h^2 \left[\int_0^1 \frac{1}{2} (\|f''(\cdot - h + sh)\|_p^p + \|f''(\cdot - h - sh)\|_p^p) ds \right]^{1/p} \\ &= h^2 \|f''\|_p. \end{aligned}$$

$\square$

In the following we shall often use the sequence spaces l_p. Some notation and results related to this spaces are necessary.

Let $a = \{a_j\}$, $j = 0, 1, \ldots$ be a sequence of real numbers, and let $1 \le p \le \infty$. Introduce the norm

$$\|a\|_{l_p} = \begin{cases} \left(\sum_{j=0}^\infty |a_j|^p \right)^{1/p}, & \text{if } 1 \le p < \infty, \\ \sup_j |a_j|, & \text{if } p = \infty. \end{cases}$$

As usually, l_p denotes the space of all sequences $a = \{a_j\}$ such that $\|a\|_{l_p} < \infty$.

We shall also need the analog of this notation for two-sided sequences $a = \{a_j\}$, $j = \ldots, -1, 0, 1, \ldots$. The space $l_p(\mathbb{Z})$ and the norm $\|a\|_{l_p}$ are defined analogously, but with the summation taken over j from $-\infty$ to ∞. Sometimes we write $\|a\|_{l_p(\mathbb{Z})}$, if it is necessary to underline the distinction between $l_p(\mathbb{Z})$ and l_p.

The following well-known lemma is the discrete analog of Lemma 8.2.

LEMMA 9.2 *Let $\{a_j\} \in l_1$ and $\{b_j\} \in l_p$ for some $1 \le p \le \infty$. Then the convolutions*

$$c_k = \sum_{m=k}^{\infty} a_m b_{m-k}, \quad c_k' = \sum_{m=0}^{k} a_{k-m} b_m$$

satisfy $\{c_k\} \in l_p$, $\{c_k'\} \in l_p$.

Let $1 \le q \le \infty$ be given, and let the function $\varepsilon(t)$ on $[0, \infty)$ be such that $\|\varepsilon\|_q^* < \infty$, where

$$\|\varepsilon\|_q^* = \begin{cases} \left(\int_0^{\infty} |\varepsilon(t)|^q \frac{dt}{t} \right)^{1/q}, & \text{if } 1 \le q < \infty, \\ \operatorname{ess\,sup}_t |\varepsilon(t)|, & \text{if } q = \infty. \end{cases}$$

Clearly, $\|\cdot\|_q^*$ is a norm in the weighted L_q-space $L_q\left([0,\infty), \frac{dt}{t}\right)$, if $q < \infty$.

DEFINITION 9.2 *Let $1 \le p, q \le \infty$ and $s = n + \alpha$, with $n \in \{0, 1, \ldots\}$, and $0 < \alpha \le 1$. The Besov space $B_p^{sq}(\mathbb{R})$ is the space of all functions f such that*

$$f \in W_p^n(\mathbb{R}) \quad \text{and} \quad \omega_p^2(f^{(n)}, t) = \varepsilon(t) t^{\alpha},$$

where $\|\varepsilon\|_q^* < \infty$.

The space $B_p^{sq}(\mathbb{R})$ is equipped with the norm

$$\|f\|_{spq} = \|f\|_{W_p^n} + \left\| \frac{\omega_p^2(f^{(n)}, t)}{t^{\alpha}} \right\|_q^*.$$

REMARK 9.1 Let us recall the Hardy inequality (DeVore & Lorentz 1993, p.24): if $\Phi \ge 0$, $\theta > 0$, $1 \le q < \infty$, then

$$\int_0^{\infty} \left\{ t^{\theta} \int_t^{\infty} \Phi(s) \frac{ds}{s} \right\}^q \frac{dt}{t} \le \frac{1}{\theta^q} \int_0^{\infty} \left\{ t^{\theta} \Phi(t) \right\}^q \frac{dt}{t}$$

and if $q = \infty$

$$\sup_{t>0}\left\{t^\theta \int_t^\infty \Phi(s)\frac{ds}{s}\right\} \leq \frac{1}{\theta}\text{ess}\sup_{t>0}\left\{t^\theta\Phi(t)\right\}.$$

Thus, if $0 < \alpha < 1$ (but *not* if $\alpha = 1$) using the Marchaud inequality we have, for $q < \infty$,

$$\int_0^\infty \left\{\frac{\omega_p^1(f^{(n)},t)}{t^\alpha}\right\}^q \frac{dt}{t} \leq \frac{1}{(1-\alpha)^q}\int_0^\infty \left\{\frac{\omega_p^2(f^{(n)},t)}{t^\alpha}\right\}^q \frac{dt}{t}$$

and, for $q = \infty$,

$$\left\|\frac{\omega_p^1(f^{(n)},t)}{t^\alpha}\right\|_\infty^* \leq \frac{1}{1-\alpha}\left\|\frac{\omega_p^2(f^{(n)},t)}{t^\alpha}\right\|_\infty^*.$$

Hence, if $0 < \alpha < 1$, we can use ω_p^1, instead of ω_p^2 in the definition of Besov spaces.

But this is not the case if $\alpha = 1$. For instance , see DeVore & Lorentz (1993, p.52), the function

$$\begin{aligned} f(x) &= x\log|x| \quad \text{if } |x| \leq 1,\\ &= 0 \qquad\qquad \text{if } |x| \geq 1, \end{aligned}$$

belongs to $B_\infty^{1\infty}$ (called also Zygmund space), but $\left\|\frac{\omega_\infty^1(f,t)}{t}\right\|_\infty^* = +\infty$. An interesting feature of this example is the following: the function f satisfies the Hölder condition of order $1 - \varepsilon$ for all $\varepsilon \in (0,1)$, but not the Hölder condition of order 1 (Lipschitz condition). This may be interpreted as the fact that the "true" regularity of f is 1, but the Hölder scale is not flexible enough to feel it. On the other hand, the scale of Besov spaces yields this opportunity.

Another example of similar kind is provided by the sample paths of the classical Brownian motion. They satisfy almost surely the Hölder condition of order α for any $\alpha < \frac{1}{2}$, but they are not $\frac{1}{2}$-Hölderian. Their "true" regularity is, however, $\frac{1}{2}$ since it can be proved that they belong to $B_p^{\frac{1}{2}\infty}$ (for any $1 \leq p < \infty$).

Definition 9.2 can be discretized, leading to the next one.

DEFINITION 9.3 *The Besov space $B_p^{sq}(I\!R)$ is the space of all functions f such that*

$$f \in W_p^n(I\!R) \text{ and } \{2^{j\alpha}\omega_p^2(f^{(n)}, 2^{-j}), \ j \in Z\!\!\!Z\} \in l_q(Z\!\!\!Z).$$

The equivalent norm of $B_p^{sq}(I\!R)$ in the discretized version is

$$\|f\|_{W_p^n} + \|\{2^{j\alpha}\omega_p^2(f^{(n)}, 2^{-j})\}\|_{l_q(Z\!\!\!Z)}.$$

The equivalence of Definitions 9.2 and 9.3 is due to the fact that the function $\omega_p^2(f^{(n)}, t)$ is non-decreasing in t, while $\frac{1}{t^\alpha}$ is decreasing. In fact,

$$\int_0^\infty \left|\frac{\omega_p^2(f^{(n)}, t)}{t^\alpha}\right|^q \frac{dt}{t} = \sum_{j=-\infty}^\infty \int_{2^j}^{2^{j+1}} \left|\frac{\omega_p^2(f^{(n)}, t)}{t^\alpha}\right|^q \frac{dt}{t},$$

and

$$\log(2)\left|\frac{\omega_p^2(f^{(n)}, 2^j)}{2^{(j+1)\alpha}}\right|^q \leq \int_{2^j}^{2^{j+1}} \left|\frac{\omega_p^2(f^{(n)}, t)}{t^\alpha}\right|^q \frac{dt}{t} \leq \log(2)\left|\frac{\omega_p^2(f^{(n)}, 2^{(j+1)})}{2^{j\alpha}}\right|^q.$$

REMARK 9.2 Using Lemma 9.2 we note that, if $0 < \alpha < 1$, one can replace $\omega_p^2(f^{(n)}, t)$ by $\omega_p^1(f^{(n)}, t)$ in the definition of $B_p^{sq}(I\!R)$. On the contrary, when s is an integer, it becomes fundamental to use $\omega_p^2(f^{(n)}, t)$. Let us observe, for instance, that $f \in L_p, \omega_p^1(f, t) = o(t)$ implies that f is constant.

9.3   Littlewood-Paley decomposition

In this section we give a characterization of Besov spaces via the Littlewood-Paley decomposition. Here we used some knowledge of the Schwartz distribution theory.

Denote $\mathcal{D}(I\!R)$ the space of infinitely many times differentiable compactly supported functions, and $S'(I\!R)$ the usual Schwartz space (the space of infinitely many times differentiable functions such that the function and all their derivatives are decreasing to zero at infinity faster than any polynomial).

Let γ be a function with the Fourier transform $\hat{\gamma}$ satisfying

- $\hat{\gamma}(\xi) \in \mathcal{D}(I\!R)$,

- supp $\hat{\gamma} \subset [-A, +A]$, $A > 0$,

- $\hat{\gamma}(\xi) = 1$ for $\xi \in \left[-\dfrac{3A}{4}, \dfrac{3A}{4} \right]$.

Let the function β be such that its Fourier transform $\hat{\beta}$ is given by

$$\hat{\beta}(\xi) = \hat{\gamma}\left(\frac{\xi}{2}\right) - \hat{\gamma}(\xi).$$

Set $\beta_j(x) = 2^j \beta(2^j x)$, $j = 0, 1, \ldots$. Note that $\hat{\beta}_j(\xi) = \hat{\beta}\left(\frac{\xi}{2^j}\right)$, and

$$\hat{\gamma}(\xi) + \sum_{j=0}^{\infty} \hat{\beta}\left(\frac{\xi}{2^j}\right) = 1. \tag{9.1}$$

Figure 9.1 presents a typical example of the Fourier transforms $\hat{\gamma}$ and $\hat{\beta}$. It follows from (9.1) that for every $f \in S'(\mathbb{R})$

$$\hat{f}(\xi) = \hat{\gamma}(\xi)\hat{f}(\xi) + \sum_{j=0}^{\infty} \hat{\beta}\left(\frac{\xi}{2^j}\right)\hat{f}(\xi). \tag{9.2}$$

This relation can be written in a different form. Define $\mathcal{D}_j f = \beta_j * f$, $j = 0, 1, \ldots$, and $\mathcal{D}_{-1} f = \gamma * f$. Then (9.2) is equivalent to

$$f = \sum_{j=-1}^{\infty} \mathcal{D}_j f \quad \text{(weakly)}, \tag{9.3}$$

or

$$\left(f - \sum_{j=-1}^{\infty} \mathcal{D}_j f, g \right) = 0, \ \forall g \in \mathcal{D}(\mathbb{R}), \tag{9.4}$$

where $(\cdot, \cdot)$ is the scalar product in $L_2(\mathbb{R})$.

The relations (9.2), (9.3) or (9.4) are called Littlewood-Paley decomposition of f.

In the following we need two lemmas.

LEMMA 9.3 (Bernstein's theorem.) *Let $f \in L_p(\mathbb{R})$, for some $1 \leq p \leq \infty$, and let the Fourier transform $\hat{f}$ satisfy: supp $\hat{f} \subset [-R, R]$, for some $R > 0$. Then there exists a constant $C > 0$ such that $\|f^{(n)}\|_p \leq C R^n \|f\|_p$, $n = 1, 2, \ldots$*

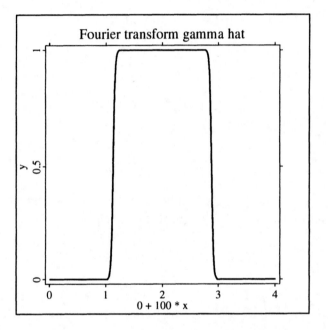

Figure 9.1: Typical example of the Fourier transforms $\hat{\gamma}$, $A = 1$.

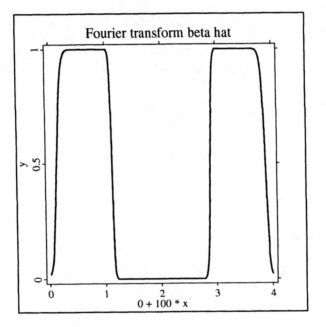

Figure 9.2: Fourier transform $\hat{\beta}$, $A = 1$.

Here is a quick proof of this lemma. Consider the function γ with $A = 2$, and let $\gamma^*(x) = R\gamma(Rx)$. Clearly, $\hat{\gamma}^*(\xi) = \hat{\gamma}(\frac{\xi}{R})$, and under the assumptions of Lemma 9.3, we have $\hat{f}(\xi) = \hat{\gamma}^*(\xi)\hat{f}(\xi)$, and hence $f = f * \gamma^*$. Therefore, $f^{(n)} = f * (\gamma^*)^{(n)}$, and in view of Lemma 8.2, $||f * (\gamma^*)^{(n)}||_p \leq R^n C||f||_p$, where $C = ||\gamma^{(n)}||_1$.

LEMMA 9.4 *Let $f \in L_p(\mathbb{R})$, $1 \leq p \leq \infty$, be such that*

$$\sum_{j=-1}^{\infty} ||\mathcal{D}_j f^{(n)}||_p < \infty,$$

for some integer $n \geq 0$ and some $1 \leq p \leq \infty$. Then $f^{(n)} \in L_p(\mathbb{R})$, and

$$\omega_p^2(f^{(n)}, t) \leq \sum_{j=-1}^{\infty} \omega_p^2(\mathcal{D}_j f^{(n)}, t), \; \forall t > 0. \tag{9.5}$$

Proof The Littlewood-Paley decomposition for $f^{(n)}$ implies that

$$||f^{(n)}||_p \leq \sum_{j=-1}^{\infty} ||\mathcal{D}_j f^{(n)}||_p < \infty.$$

Hence, $f^{(n)} \in L_p(\mathbb{R})$. Quite similarly,

$$||\Delta_h^2 f^{(n)}||_p \leq \sum_{j=-1}^{\infty} ||\Delta_h^2 \mathcal{D}_j f^{(n)}||_p < \infty,$$

for any $h > 0$. By Lemma 9.1 (i) we have also $\omega_p^2(\mathcal{D}_j f^{(n)}, t) < \infty$, $\forall j = -1, 0, \ldots$. Combining these facts with the observation that $\omega_p^2(f + g, t) \leq \omega_p^2(f, t) + \omega_p^2(g, t)$, for any functions f, g, we get (9.5). $\quad\square$

THEOREM 9.1 *If $1 \leq p, q \leq \infty$, $s > 0$, and $f \in L_p(\mathbb{R})$, we have :* $f \in B_p^{sq}(\mathbb{R})$ *if and only if*

$$||\mathcal{D}_{-1} f||_p < \infty \text{ and } \left\{ 2^{js} ||\mathcal{D}_j f||_p, \; j = 0, 1, \ldots \right\} \in l_q. \tag{9.6}$$

Proof

Necessity of (9.6). Assume that $f \in B_p^{sq}(\mathbb{R}), s = n + \alpha, 0 < \alpha \leq 1$, and let us prove (9.6). Clearly, the function $\hat{\beta}\left(\frac{\xi}{2^j}\right)\hat{f}(\xi)$ is compactly supported, and in view of (4.10), we have

$$(i\xi)^n \hat{\beta}\left(\frac{\xi}{2^j}\right)\hat{f}(\xi) = \mathcal{F}[(\beta_j * f)^{(n)}](\xi).$$

Hence,

$$\begin{aligned} \hat{\beta}\left(\frac{\xi}{2^j}\right)\hat{f}(\xi) &= 2^{-jn}\left(\frac{2^j}{i\xi}\right)^n \hat{\beta}\left(\frac{\xi}{2^j}\right)\hat{f}(\xi)(i\xi)^n \\ &= 2^{-jn}(-i)^n \hat{\gamma}_n\left(\frac{\xi}{2^j}\right)\mathcal{F}[(\beta_j * f)^{(n)}](\xi), \end{aligned}$$

where $\hat{\gamma}_n$ is a function of $\mathcal{D}(\mathbb{R})$ defined by : $\hat{\gamma}_n(\xi) = \frac{\hat{\delta}(\xi)}{\xi^n}$, and $\hat{\delta}$ is a function from $\mathcal{D}(\mathbb{R})$ which equals 1 on the support of $\hat{\beta}$ and 0 in a neighborhood of 0. Hence, by Lemma 8.2,

$$\|\mathcal{D}_j f\|_p \leq \|\gamma_n\|_1 2^{-jn}\|(\beta_j * f)^{(n)}\|_p = \|\gamma_n\|_1 2^{-jn}\|\beta_j * f^{(n)}\|_p, \quad j = 0, 1, \ldots, \tag{9.7}$$

where γ_n is the inverse Fourier transform of $\hat{\gamma}_n$. The last equality in (9.7) is justified by the use of partial integration and by the fact that $\|\beta_j * f^{(n)}\|_p < \infty$ shown below. Let us evaluate $\|\beta_j * f^{(n)}\|_p$. We have $\int \beta_j(y)dy = 0$, since $\hat{\beta}_j(0) = 0$, and also β_j is an even function. Thus,

$$\begin{aligned} \beta_j * f^{(n)}(x) &= \int \beta_j(y)f^{(n)}(x - y)dy \\ &= \frac{1}{2}\int \beta_j(y)\left\{f^{(n)}(x - y) - 2f^{(n)}(x) + f^{(n)}(x + y)\right\}dy \\ &= \frac{1}{2}\int \beta(y)\left\{f^{(n)}(x - 2^{-j}y) - 2f^{(n)}(x) + f^{(n)}(x + 2^{-j}y)\right\}dy, \end{aligned}$$

and, by Lemma 8.1 and Lemma 9.1 (iv),

$$\begin{aligned} \|\beta_j * f^{(n)}\|_p &\leq \int |\beta(y)|\omega_p^2(f^{(n)}, 2^{-j}|y|)dy \\ &\leq \omega_p^2(f^{(n)}, 2^{-j})\int |\beta(y)|(1 + |y|)^2 dy \\ &\leq C_1 \omega_p^2(f^{(n)}, 2^{-j}), \end{aligned} \tag{9.8}$$

where C_1 is a positive constant (the last integral is finite: in fact, since $\hat{\beta}$ is infinitely many times differentiable and compactly supported, the function β is uniformly bounded and, by Lemma 4.1 of Chapter 4, $|\beta(x)||x|^N \to 0$, as $|x| \to \infty$, for any $N \geq 1$).

From (9.7) and (9.8) we deduce

$$
\begin{aligned}
2^{js}||\mathcal{D}_j f||_p &\leq C_1 ||\gamma_n||_1 2^{j(s-n)} \omega_p^2(f^{(n)}, 2^{-j}) \\
&= C_2 2^{j\alpha} \omega_p^2(f^{(n)}, 2^{-j}),
\end{aligned}
\tag{9.9}
$$

where $C_2 > 0$ is a constant. By Definition 9.3, if $f \in B_p^{sq}(I\!R)$, then $\{2^{j\alpha}\omega_p^2(f^{(n)}, 2^{-j})\} \in l_q(Z\!\!Z)$. This and (9.9) yield: $\{2^{js}||\mathcal{D}_j f||, j = 0, 1, \ldots\} \in l_q$. The inequality $||\mathcal{D}_{-1}f||_p < \infty$ is straightforward.

Sufficiency of (9.6). Suppose that $||\mathcal{D}_{-1}f||_p < \infty$, $||\mathcal{D}_j f||_p = 2^{-js}\eta_j$, $j = 0, 1, \ldots$, where $\{\eta_j\} \in l_q$, and let us show that $f \in B_p^{sq}(I\!R)$. We have

$$
\mathcal{F}[(\beta_j * f)^{(n)}](\xi) = (i\xi)^n \hat{\beta}\left(\frac{\xi}{2^j}\right) \hat{f}(\xi) = i^n \hat{\gamma}_{-n}\left(\frac{\xi}{2^j}\right) 2^{jn} \hat{f}(\xi) \hat{\beta}\left(\frac{\xi}{2^j}\right)
\tag{9.10}
$$

Lemma 8.2 and (9.10) entail:

$$
\begin{aligned}
||\mathcal{D}_j f^{(n)}||_p &\leq 2^{jn} ||2^j \gamma_{-n}(2^j \cdot)||_1 \cdot ||\mathcal{D}_j f||_p \\
&= ||\gamma_{-n}||_1 \eta_j 2^{-j\alpha}, \quad j \geq 0
\end{aligned}
\tag{9.11}
$$

This yields, in particular, that

$$
\sum_{j=-1}^{\infty} ||\mathcal{D}_j f^{(n)}||_p < \infty,
\tag{9.12}
$$

and, by Lemma 9.4, $f^{(n)} \in L_p(I\!R)$.

Using the Definition 9.3, it remains to prove that $\{2^{k\alpha}\omega_p^2(f^{(n)}, 2^{-k}), k \in Z\!\!Z\} \in l_q(Z\!\!Z)$. For $k < 0$ we use the rough estimate from Lemma 9.1 (i):

$$
2^{k\alpha}\omega_p^2(f^{(n)}, 2^{-k}) \leq 4||f^{(n)}||_p 2^{k\alpha} = C_3 2^{k\alpha},
$$

where $C_3 > 0$ is a constant. This entails

$$
\sum_{k=-\infty}^{-1} \left\{2^{k\alpha}\omega_p^2(f^{(n)}, 2^{-k})\right\}^q \leq C_3^q \sum_{k=1}^{\infty} 2^{-kq\alpha} < \infty, 1 \leq q < \infty,
\tag{9.13}
$$

and

$$\max_{-\infty \le k \le -1} 2^{k\alpha} \omega_p^2(f^{(n)}, 2^{-k}) < \infty, \tag{9.14}$$

for $q = \infty$.

For $k \ge 0$, the evaluation is more delicate. Note that the support of the Fourier transform $\mathcal{F}[\mathcal{D}_j f^{(n)}]$ is included in the interval $[-2^{j+1}A, 2^{j+1}A])$, and thus, by Lemma 9.3,

$$\|(\mathcal{D}_j f^{(n)})''\|_p \le C_4 2^{-2j} \|\mathcal{D}_j f^{(n)}\|_p, \tag{9.15}$$

where $C_4 > 0$ is a constant, $j \ge -1$. Using Lemma 9.1 (vi), (9.11) and (9.15), we find

$$
\begin{aligned}
\omega_p^2(\mathcal{D}_j f^{(n)}, 2^{-k}) &\le 2^{-2k} \|(\mathcal{D}_j f^{(n)})''\|_p \le C_4 \|\gamma_{-n}\|_1 2^{-2(k+j+\alpha)} \eta_j \\
&\le C_5 2^{-(k+j)} \eta_j 2^{-k\alpha}, \; j \ge 0, \; k \ge 0, \tag{9.16}
\end{aligned}
$$

where $C_5 > 0$ is a constant.

Recalling (9.12) and using Lemma 9.4, we get, for any $k \ge 0$,

$$
\begin{aligned}
\omega_p^2(f^{(n)}, 2^{-k}) &\le \sum_{j=-1}^{\infty} \omega_p^2(\mathcal{D}_j f^{(n)}, 2^{-k}) \\
&= \omega_p^2(\mathcal{D}_{-1} f^{(n)}, 2^{-k}) + \sum_{j=0}^{k-1} \omega_p^2(\mathcal{D}_j f^{(n)}, 2^{-k}) \\
&\quad + \sum_{j=k}^{\infty} \omega_p^2(\mathcal{D}_j f^{(n)}, 2^{-k}). \tag{9.17}
\end{aligned}
$$

Here, in view of (9.16),

$$
\begin{aligned}
\sum_{j=0}^{k-1} \omega_p^2(\mathcal{D}_j f^{(n)}, 2^{-k}) &\le C_5 2^{-k\alpha} \sum_{j=0}^{k-1} 2^{-(k+j)} \eta_j \\
&\le 2^{-k\alpha} \eta'_k, \; \eta'_k = C_5 \sum_{m=k}^{\infty} 2^{-m} \eta_{m-k}. \tag{9.18}
\end{aligned}
$$

Here $\{\eta'_k\} \in l_q$ by Lemma 9.2.

On the other hand, by Lemma 9.1 (i) and (9.11),

$$
\begin{aligned}
\sum_{j=k}^{\infty} \omega_p^2(\mathcal{D}_j f^{(n)}, 2^{-k}) &\le 4 \sum_{j=k}^{\infty} \|\gamma_{-n}\|_1 \eta_j 2^{-j\alpha} \\
&= 4\|\gamma_{-n}\|_1 2^{-k\alpha} \sum_{j=k}^{\infty} \eta_j 2^{-\alpha(j-k)} = \tilde{\eta}_k 2^{-k\alpha}, \tag{9.19}
\end{aligned}
$$

where again $\{\tilde{\eta}_k\} \in l_q$ by Lemma 9.2.

Finally, the same reasoning as in (9.11), (9.15) and (9.16) yields

$$\omega_p^2(\mathcal{D}_{-1}f^{(n)}, 2^{-k}) \leq 2^{-2k} \|(\mathcal{D}_{-1}f^{(n)})''\|_p$$
$$\leq C_6 2^{-2k} \|\mathcal{D}_{-1}f^{(n)}\|_p \leq C_7 2^{-2k}, \tag{9.20}$$

where we used (9.12). Here C_6 and C_7 are positive constants.

To finish the proof, it remains to put together (9.17) – (9.20), which yields

$$\{2^{k\alpha}\omega_p^2(f^{(n)}, 2^{-k}), \ k = 0, 1, \ldots\} \in l_q,$$

and to combine this with (9.13) and (9.14). Thus, finally

$$\{2^{k\alpha}\omega_p^2(f^{(n)}, 2^{-k}), \ k \in \mathbb{Z}\} \in l_q(\mathbb{Z}),$$

and the theorem is proved.   □

Theorem 9.1 allows to obtain the following characterization of Besov spaces.

THEOREM 9.2 (Characterization of Besov spaces.) *Let $N \geq 0$ be an integer, let $0 < s < N + 1$, $1 \leq p, q \leq \infty$, and let f be a Borel function on $\mathbb{R}$. The necessary and sufficient condition for $f \in B_p^{sq}(\mathbb{R})$ is*

$$f = \sum_{j=0}^{\infty} u_j \quad (weakly), \tag{9.21}$$

where the functions u_j satisfy

$$\|u_j\|_p \leq 2^{-js}\varepsilon_j, \ \|u_j^{(N+1)}\|_p \leq 2^{j(N+1-s)}\varepsilon_j' \tag{9.22}$$

with $\{\varepsilon_j\} \in l_q$, $\{\varepsilon_j'\} \in l_q$.

REMARK 9.3 Equality (9.21) is assumed to hold in the same sense as the Littlewood-Paley decomposition. Namely, $(f - \sum_{j=0}^{\infty} u_j, g) = 0$, $\forall g \in \mathcal{D}(\mathbb{R})$, is an equivalent version of (9.21).

Proof of Theorem 9.2 The necessary part is a direct consequence of Theorem 9.1, if one takes $u_j = \mathcal{D}_{j-1}f$. The second inequality in (9.22) follows then from Lemma 9.3 (in fact, the support of the Fourier transform $\mathcal{F}[\mathcal{D}_j f]$ is included in the interval $[-2^{j+1}A, 2^{j+1}A]$).

Let us prove that conditions (9.21) and (9.22) are sufficient for $f \in B_p^{sq}(\mathbb{R})$. Under these conditions we have

$$||\mathcal{D}_j u_m||_p \le ||\beta||_1 ||u_m||_p \le ||\beta||_1 2^{-ms} \varepsilon_m,$$

for any integers $j \ge -1, m \ge 0$. Therefore, the series $\sum_{m=0}^{\infty} \mathcal{D}_j u_m$ converges in $L_p(\mathbb{R})$, and

$$|| \sum_{m=j}^{\infty} \mathcal{D}_j u_m ||_p \le \sum_{m=j}^{\infty} ||\mathcal{D}_j u_m||_p$$

$$\le ||\beta||_1 2^{-js} \sum_{m=j}^{\infty} 2^{-(m-j)s} \varepsilon_m = 2^{-js} \eta_j, \tag{9.23}$$

where $\{\eta_j\} \in l_q$ by Lemma 9.2.

Now,

$$\mathcal{D}_j f = \sum_{m=0}^{j-1} \mathcal{D}_j u_m + \sum_{m=j}^{\infty} \mathcal{D}_j u_m. \tag{9.24}$$

Let us evaluate the first sum in (9.24). Note that the Fourier transform

$$\mathcal{F}[\mathcal{D}_j u_m](\xi) = \hat{\beta}\left(\frac{\xi}{2^j}\right) 2^{-j(N+1)} (i\xi)^{N+1} \hat{u}_m(\xi)(-i)^{N+1} \left(\frac{2^j}{\xi}\right)^{N+1}$$

$$= (-i)^{N+1} 2^{-j(N+1)} \mathcal{F}[u_m^{(N+1)}](\xi) \hat{\beta}\left(\frac{\xi}{2^j}\right) \hat{\gamma}_{N+1}\left(\frac{\xi}{2^j}\right) \tag{9.25}$$

where as in the proof of Theorem 9.1, $\hat{\gamma}_{N+1} \in \mathcal{D}(\mathbb{R})$ is a function defined by $\hat{\gamma}_{N+1}(\xi) = \hat{\delta}(\xi)/\xi^{N+1}$ with $\hat{\delta} \in \mathcal{D}(\mathbb{R})$ that equals 1 on the support of $\hat{\beta}$, and 0 in a neighborhood of 0. Taking the inverse Fourier transforms of both sides of (9.25) and applying Lemma 8.2 and (9.22), we obtain

$$||\mathcal{D}_j u_m||_p \le 2^{-j(N+1)} ||u_m^{(N+1)}||_p ||\beta_j * 2^j \gamma_{N+1}(2^j \cdot)||_1$$
$$\le 2^{-j(N+1)} ||\beta_j||_1 ||\gamma_{N+1}||_1 2^{m(N+1-s)} \varepsilon_m'.$$

This implies

$$|| \sum_{m=0}^{j-1} \mathcal{D}_j u_m ||_p \le \sum_{m=0}^{j-1} ||\mathcal{D}_j u_m||_p$$

$$\le C_8 2^{-js} \sum_{m=0}^{j} \varepsilon_m' 2^{(m-j)(N+1-s)} \le 2^{-js} \eta_j', \tag{9.26}$$

where $\{\eta_j'\} \in l_p$ by Lemma 9.2.

Putting together (9.23), (9.24) and (9.26), we get (9.6), and thus $f \in B_p^{sq}(I\!\!R)$ by Theorem 9.1.   □

9.4   Approximation theorem in Besov spaces

Here and later in this chapter we use the approximation kernels K and refer to the Conditions $M(N), H(N)$, introduced in Section 8.3.

The result of this section is an analog of Theorem 8.1 (ii) for the Besov spaces.

THEOREM 9.3 *Let the kernel K satisfy the Condition $M(N)$, and Condition $H(N+1)$ for some integer $N \geq 0$. Let $1 \leq p, q \leq \infty$ and $0 < s < N+1$. If $f \in B_p^{sq}(I\!\!R)$, then*

$$\|K_j f - f\|_p = 2^{-js} \varepsilon_j,$$

where $\{\varepsilon_j\} \in l_q$.

Proof   Let $\sum_{k=-1}^{\infty} g_k$, where $g_k = \mathcal{D}_k f$, be the Littlewood-Paley decomposition of f. Then, clearly, $K_j f - f$ has the Littlewood-Paley decomposition $\sum_{k=-1}^{\infty} (K_j g_k - g_k)$, and

$$
\begin{aligned}
\|K_j f - f\|_p &\leq \sum_{k=-1}^{\infty} \|K_j g_k - g_k\|_p \\
&\leq \sum_{k=-1}^{j} \|K_j g_k - g_k\|_p + \sum_{k=j+1}^{\infty} (\|F\|_1 + 1)\|g_k\|_p, \quad (9.27)
\end{aligned}
$$

where the Condition $H(N+1)$ and Lemma 8.2 were used. By Theorem 9.1,

$$\|g_k\|_p = 2^{-ks} \varepsilon_k', \quad \{\varepsilon_k'\} \in l_q. \tag{9.28}$$

Note that the support of the Fourier transform $\mathcal{F}[g_k]$ is included in $[-2^{k+1}A, 2^{k+1}A]$, and thus, by virtue of Lemma 9.3,

$$\|g_k^{(N+1)}\|_p \leq C_9 2^{(N+1)k}\|g_k\|_p \leq C_9 2^{(N+1-s)k} \varepsilon_k', \tag{9.29}$$

where $C_9 > 0$ is a constant. Thus, g_k satisfies the assumptions of Theorem 8.1 (ii). Acting as in the proof of Theorem 8.1 and using Lemma 9.1 (iii), we obtain, for any $h > 0$,

$$
\begin{aligned}
\|K_h g_k - g_k\|_p &\leq h^N \int_0^1 du \int_{-\infty}^\infty |t|^N F(t)\frac{(1-u)^{N-1}}{(N-1)!}\|\tau_{-tuh}(g_k^{(N)}) - g_k^{(N)}\|_p dt \\
&\leq h^N \omega_p^1(g_k^{(N)}, h) \int_0^1 du \int_{-\infty}^\infty F(t)(1+|ut|)|t|^N\frac{(1-u)^{N-1}}{(N-1)!}dt \\
&\leq C_{10} h^N \omega_p^1(g_k^{(N)}, h), \qquad\qquad (9.30)
\end{aligned}
$$

where $C_{10} > 0$ is a constant that does not depend on h. Set $h = 2^{-j}$. Then, by Lemma 9.1 (v), (9.29) and (9.30),

$$
\|K_j g_k - g_k\|_p \leq 2C_9 C_{10} 2^{-(N+1)j+(N+1-s)k}\varepsilon_k'. \qquad (9.31)
$$

Using (9.28) and (9.31), we can reduce (9.27) to the form

$$
\|K_j f - f\|_p \leq C_{11} 2^{-js}\Big[\sum_{k=-1}^j 2^{-(N+1-s)(j-k)}\varepsilon_k' + \sum_{k=j+1}^\infty \varepsilon_k' 2^{-(k-j)s}\Big] = 2^{-js}\varepsilon_j,
$$

where $\{\varepsilon_j\} \in l_q$ by Lemma 9.2. □

REMARK 9.4 As in (9.30) we can obtain directly for f the following inequality:

$$
\|K_h f - f\|_p \leq C_{10} h^N \omega_p^1(f^{(N)}, h),
$$

which yields immediately Theorem 9.3 for the case where $s = n + \alpha$ with $0 < \alpha < 1$ (using Remark 9.2). The necessity of the Littlewood-Paley decomposition is the price to pay to cover the case of integer s as well.

9.5 Wavelets and approximation in Besov spaces

Here we show that under certain general conditions the wavelet expansion is analogous to the Littlewood-Paley decomposition. This yields the characterization of Besov spaces in terms of wavelet coefficients.

Let φ be the scaling function of a multiresolution analysis (a father wavelet). Let, as always, $\varphi_k(x) = \varphi(x - k)$, $k \in \mathbb{Z}$,

$$
\begin{aligned}
\varphi_{jk}(x) &= 2^{j/2}\varphi(2^j x - k), \\
\psi_{jk}(x) &= 2^{j/2}\psi(2^j x - k), \ k \in \mathbb{Z}, \ j = 0, 1, \ldots,
\end{aligned}
$$

where ψ is the associated mother wavelet.

As follows from the results of Chapters 5 and 8, under rather general conditions on φ, any function $f \in L_p(\mathbb{R})$, $p \in [1, \infty)$, has the following expansion

$$f(x) = \sum_k \alpha_k \varphi_k(x) + \sum_{j=0}^{\infty} \sum_k \beta_{jk} \psi_{jk}(x), \qquad (9.32)$$

where the series converges in $L_p(\mathbb{R})$, and

$$\alpha_k = \int \overline{\varphi(x-k)} f(x) dx,$$

$$\beta_{jk} = 2^{j/2} \int \overline{\psi(2^j x - k)} f(x) dx.$$

Consider the associated kernel

$$K_j(x, y) = 2^j \sum_k \varphi(2^j x - k) \overline{\varphi(2^j y - k)}.$$

Using the notation of Section 8.3, we can write, for any function $f \in L_p(\mathbb{R})$ and any integer j,

$$K_j f(x) = \sum_k \alpha_{jk} 2^{j/2} \varphi(2^j x - k) = \sum_k \alpha_{jk} \varphi_{jk}(x),$$

$$K_j f(x) = \sum_k \alpha_k \varphi_k(x) + \sum_{m=0}^{j-1} \sum_k \beta_{mk} \psi_{mk}(x),$$

where K_j is the orthogonal projection operator on the space V_j spanned by $\{\varphi_{jk}, k \in \mathbb{Z}\}$ and as usual

$$\alpha_{jk} = 2^{j/2} \int \overline{\varphi(2^j x - k)} f(x) dx.$$

Thus,

$$K_{j+1} f(x) - K_j f(x) = \sum_k \beta_{jk} \psi_{jk}(x). \qquad (9.33)$$

Let $||\alpha_j||_{l_p}$ be the $l_p(\mathbb{Z})$-norm of the sequence $\{\alpha_{jk}, k \in \mathbb{Z}\}$, for a fixed $j \in \{0, 1, \ldots\}$. Suppose that φ satisfies the Condition (θ) introduced in Section 8.5. Then, by Proposition 8.6 (v), Condition (θ) is true for the mother wavelet ψ as well.

Applying Proposition 8.3, we get that there exist two positive constants, C_{12} and C_{13}, such that

$$C_{12}2^{j(\frac{1}{2}-\frac{1}{p})}||\alpha_j||_{l_p} \leq ||K_jf||_p \leq C_{13}2^{j(\frac{1}{2}-\frac{1}{p})}||\alpha_j||_{l_p}, \tag{9.34}$$

$$C_{12}2^{j(\frac{1}{2}-\frac{1}{p})}||\beta_j||_{l_p} \leq ||K_{j+1}f - K_jf||_p \leq C_{13}2^{j(\frac{1}{2}-\frac{1}{p})}||\beta_j||_{l_p}, \tag{9.35}$$

for any integer $j \geq 0$.

THEOREM 9.4 *Let φ be a scaling function, satisfying (8.33), (8.34) and the Condition $S(N+1)$, for some integer $N \geq 0$. Let, in addition, φ satisfy one of the conditions (W1) to (W4) of Corollary 8.2 (ensuring the Condition $M(N)$). Then, for any $0 < s < N+1$, and $1 \leq p,q \leq \infty$ we have:*

(i) *$f \in B_p^{sq}(I\!R) \Longrightarrow$*
   $f \in L_p(I\!R)$ and $||K_jf - f||_p = 2^{-js}\varepsilon_j$, $j = 0,1,\ldots$, with $\{\varepsilon_j\} \in l_q$,

(ii) *$f \in B_p^{sq}(I\!R) \Longrightarrow$*
   $||\alpha_0||_{l_p} < \infty$ and $||\beta_j||_{l_p} = 2^{-j(s+\frac{1}{2}-\frac{1}{p})}\varepsilon_j'$, $j = 0,1,\ldots$, with $\{\varepsilon_j'\} \in l_q$.

Proof

(i) This is a direct consequence of Theorem 9.3, since Condition $S(N+1)$ implies Condition $H(N+1)$.

(ii) From (9.34) and Remark 8.3 we get

$$||\alpha_0||_{l_p} \leq C_{12}^{-1}||Kf||_p \leq C_{12}^{-1}||F||_1||f||_p < \infty.$$

On the other hand, (9.35) and part (i) of the present theorem entail

$$C_{12}2^{j(\frac{1}{2}-\frac{1}{p})}||\beta_j||_{l_p} \leq ||K_{j+1}f - f||_p + ||f - K_jf||_p$$
$$\leq 2^{-js}(\varepsilon_j + \frac{1}{2}\varepsilon_{j+1}) = 2^{-js}\varepsilon_j',$$

where $\{\varepsilon_j'\} \in l_p$.

$\square$

REMARK 9.5 A weaker result may be obtained for the case where φ is a father wavelet satisfying Condition S. Then, in view of Corollary 8.1, the kernel K satisfies Condition $M(0)$, and one can apply Theorem 8.1 (i). This yields

$$||K_j f - f||_p \to 0, \text{ as } j \to \infty, \qquad (9.36)$$

if either $1 \leq p < \infty$, and $f \in L_p(\mathbb{R})$, or $p = \infty, f \in L_\infty(\mathbb{R})$ and f is uniformly continuous. Also

$$K_j f \to f, \text{ as } j \to \infty,$$

in the weak topology $\sigma(L_\infty, L_1)$, $\forall f \in L_\infty(\mathbb{R})$. In fact, for any $g \in L_1(\mathbb{R})$ we have

$$\int g(x) K_j f(x) dx = \int f(u) \tilde{K}_j g(u) du,$$

where $\tilde{K}(u, v) = K(v, u)$. But $\tilde{K}$ satisfies also the Condition $M(0)$, so $||\tilde{K}_j g - g||_1 \to 0$, as $j \to \infty$. This implies

$$\forall g \in L_1(\mathbb{R}), \int g(x) K_j f(x) dx \to \int f(x) g(x) dx, \text{ as } j \to \infty.$$

$\square$

One can compare Theorem 9.4 with Corollary 8.2, which contains a similar result for the Sobolev spaces. Note that the assumptions on the father wavelet φ in both results are the same. Moreover, the result of Corollary 8.2 can be formulated as follows: for any $1 \leq p \leq \infty$,

$$f \in W_p^{N+1}(\mathbb{R}) \Rightarrow f \in L_p(\mathbb{R}) \text{ and } ||K_j f - f||_p = 2^{-j(N+1)} \varepsilon_j,$$

with $\{\varepsilon_j\} \in l_\infty$. This and the argument in the proof of Theorem 9.4 (ii) yield also:

$$f \in W_p^{N+1}(\mathbb{R}) \Rightarrow ||\alpha_0||_{l_p} < \infty \text{ and } ||\beta_j||_{l_p} = 2^{-j(N+\frac{3}{2}-\frac{1}{p})} \varepsilon_j', \qquad (9.37)$$

with $\{\varepsilon_j'\} \in l_\infty$. Using Theorem 8.1 (i) and Theorem 8.3 one can get that, under the assumptions of Theorem 9.4, for any $k = 0, 1, \ldots, N$,

$$f \in \tilde{W}_p^k(\mathbb{R}) \Rightarrow f \in L_p(\mathbb{R}) \text{ and } ||K_j f - f||_p = 2^{-jk} \varepsilon_j, \text{with } \{\varepsilon_j\} \in c_0,$$

and

$$f \in \tilde{W}_p^k(\mathbb{R}) \Rightarrow \quad ||\alpha_0||_{l_p} < \infty \text{ and}$$

$$||\beta_j||_{l_p} = 2^{-j(k+\frac{1}{2}-\frac{1}{p})}\varepsilon_j', \text{ with } \{\varepsilon_j'\} \in c_0. \quad (9.38)$$

Here c_0 is the space of sequences tending to 0.

It turns out that the results (9.37) and (9.38) cannot be inverted. That is, the Sobolev spaces cannot be characterized in terms of wavelet coefficients. The situation changes drastically for the Besov spaces, where such a characterization is possible. This is shown in the next two theorems.

THEOREM 9.5 *Let φ be a scaling function satisfying (8.33), (8.34) and the Condition (θ). Let $N \geq 0$ be an integer. Assume that φ is $N+1$ times weakly differentiable and that the derivative $\varphi^{(N+1)}$ satisfies the Condition (θ). Then, for any $0 < s < N+1$, $1 \leq p, q \leq \infty$, and any function $f \in L_p(\mathbb{R})$ we have*

(i) $||K_j f - f||_p = \varepsilon_j 2^{-js}$, $j = 0, 1, \ldots$, with $\{\varepsilon_j\} \in l_q \Longrightarrow f \in B_p^{sq}(\mathbb{R})$,

(ii) $(||\alpha_0||_{l_p} < \infty$ and $||\beta_j||_{l_p} = 2^{-j(s+\frac{1}{2}-\frac{1}{p})}\varepsilon_j'$, $j = 0, 1, \ldots$, with

$$\{\varepsilon_j'\} \in l_q) \Longrightarrow f \in B_p^{sq}(\mathbb{R}).$$

Proof

(i) Set $u_0 = K_0 f = Kf$, $u_j = K_{j+1}f - K_j f$. Then

$$\begin{aligned} ||u_j||_p &\leq 2^{-js}(\varepsilon_j + 2^{-1}\varepsilon_{j+1}) \\ &= 2^{-js}\eta_j, \text{ where } \{\eta_j\} \in l_q. \end{aligned} \quad (9.39)$$

Next, for some coefficients $\{\lambda_{jk}\}$ we have

$$u_j(x) = \sum_k \lambda_{jk} 2^{(j+1)/2}\varphi(2^{j+1}x - k),$$

since $K_{j+1}f - K_j f \in V_{j+1}$. Thus, by Proposition 8.3,

$$C_{12}2^{(j+1)(\frac{1}{2}-\frac{1}{p})}||\lambda_j||_{l_p} \leq ||u_j||_p \leq C_{13}2^{(j+1)(\frac{1}{2}-\frac{1}{p})}||\lambda_j||_{l_p}. \quad (9.40)$$

But
$$u_j^{(N+1)}(x) = 2^{(j+1)(N+1)} \sum_k \lambda_{jk} 2^{\frac{j+1}{2}} \varphi^{(N+1)}(2^{j+1}x - k),$$

and using the assumptions of the theorem and Proposition 8.3 we get

$$||u_j^{(N+1)}||_p \le C_{13} 2^{(j+1)(N+1)} 2^{(j+1)(\frac{1}{2} - \frac{1}{p})} ||\lambda_j||_{l_p}.$$

This, together with (9.39) and (9.40) yield

$$
\begin{aligned}
||u_j^{(N+1)}||_p &\le C_{13} C_{12}^{-1} 2^{(j+1)(N+1)} ||u_j||_p \\
&\le C_{14} 2^{j(N+1)} ||u_j||_p = C_{14} 2^{j(N+1-s)} \eta_j.
\end{aligned}
\tag{9.41}
$$

It remains to note that (9.39) and (9.41) guarantee (9.22), while (9.21) follows directly from the construction of u_j. Thus, applying Theorem 9.2, we obtain that $f \in B_p^{sq}(\mathbb{R})$.

(ii) The imposed assumptions imply, jointly with (9.34) and (9.35), that

$$||Kf||_p < \infty, \quad ||K_{j+1}f - K_j f||_p \le \varepsilon_j'' 2^{-js},$$

with $\{\varepsilon_j''\} \in l_q$. Therefore,

$$\sum_{j=0}^{\infty} ||K_{j+1}f - K_j f||_p < \infty,$$

and the series

$$Kf + \sum_{j=0}^{\infty} (K_{j+1}f - K_j f)$$

converges in $L_p(\mathbb{R})$. Its limit is f. In fact,

$$Kf + \sum_{j=0}^{j_0-1} (K_{j+1}f - K_j f) = K_{j_0} f,$$

for any integer $j_0 \ge 1$, and therefore

$$
\begin{aligned}
||K_{j_0}f &- f||_p \\
&= ||\sum_{j=j_0}^{\infty} (K_{j+1}f - K_j f)||_p \le \sum_{j=j_0}^{\infty} ||K_{j+1}f - K_j f||_p \\
&\le \sum_{j=j_0}^{\infty} \varepsilon_j'' 2^{-js} = 2^{-j_0 s} \sum_{j=j_0}^{\infty} \varepsilon_j'' 2^{-(j-j_0)s} = 2^{-j_0 s} \eta_{j_0},
\end{aligned}
$$

where $\{\eta_{j_0}\} \in l_q$ by Lemma 9.2.

To end the proof it suffices to use the part (i) of the present theorem. $\square$

THEOREM 9.6 *Let φ be a scaling function satisfying (8.33), (8.34) and the Condition $S(N+1)$, for some integer $N \geq 0$. Assume that φ is $N+1$ times weakly differentiable and that the derivative $\varphi^{(N+1)}$ satisfies the Condition (θ). Then, for any $0 < s < N+1$, $1 \leq p, q \leq \infty$, and any function $f \in L_p(\mathbb{R})$ the following conditions are equivalent:*

(B1) $f \in B_p^{sq}(\mathbb{R})$,

(B2) $\|K_j f - f\|_p = 2^{-js}\varepsilon_j$, $j = 0, 1, \ldots$, where $\{\varepsilon_j\} \in l_q$,

(B3) $\|\alpha_0\|_{l_p} < \infty$ and $\|\beta_j\|_{l_p} = 2^{-j(s+\frac{1}{2}-\frac{1}{p})}\varepsilon_j'$, $j = 0, 1, \ldots$, where $\{\varepsilon_j'\} \in l_q$.

Proof Implications (B2) $\Longrightarrow$ (B1) and (B3) $\Longrightarrow$ (B1) follow from Theorem 9.5, since Condition $S(N+1)$ implies Condition (θ) (see Lemma 8.5).

Implications (B1) $\Longrightarrow$ (B2) and (B1) $\Longrightarrow$ (B3) follow from Theorem 9.4, since under the imposed assumptions we have $\varphi \in W_1^{N+1}(\mathbb{R})$ (and thus the condition (W1) of Corollary 8.2 holds). $\square$

COROLLARY 9.1 *Under the assumptions of Theorem 9.6 the Besov norm $\|f\|_{spq}, 1 \leq p < \infty, 1 \leq q < \infty$, is equivalent to the following norm in the space of wavelet coefficients:*

$$\|f\|_{spq}' = \left(\sum_k |\alpha_k|^p\right)^{\frac{1}{p}} + \left[\sum_{j=0}^{\infty}\left(2^{j(s+\frac{1}{2}-\frac{1}{p})}\left(\sum_k |\beta_{jk}|^p\right)^{\frac{1}{p}}\right)^q\right]^{\frac{1}{q}}$$

where

$$\alpha_k = \int f(x)\overline{\varphi_k(x)}dx,$$
$$\beta_{jk} = \int f(x)\overline{\psi_{jk}(x)}dx.$$

EXAMPLE 9.1 To approximate correctly a function of $B_p^{sq}(I\!R)$ with $s <$
$N + 1$, it is sufficient to use the wavelet expansion with the Daubechies
$D2(N + 1)$ father wavelet φ, as discussed in Example 8.1.

However, the characterization of the Besov space $B_p^{sq}(I\!R)$ in terms of
wavelet expansions requires more regular wavelets. In fact, to apply Theorem 9.6, we need that φ were $N + 1$ times weakly differentiable. In view
of (7.10), within the Daubechies family, this property is ensured only for
wavelets $D12(N + 1)$ and higher, and asymptotically (if N is large enough)
for wavelets $D10(N + 1)$ and higher.

Finally, observe that certain embedding theorems can be easily obtained
using the previous material. For example, we have the following result.

COROLLARY 9.2 *Let* $s > 0,\ 1 \le p \le p' \le \infty,$ *and* $1 \le q \le q' \le \infty.$
Then

(i) $B_p^{sq}(I\!R) \subset B_p^{sq'}(I\!R),$

(ii) $B_p^{k1}(I\!R) \subset W_p^k(I\!R) \subset B_p^{k\infty}(I\!R),$ *for any integer* $k > 0,$

(iii) $B_p^{sq}(I\!R) \subset B_{p'}^{s'q}(I\!R),$ *if* $s' - \dfrac{1}{p'} = s - \dfrac{1}{p},$

(iv) $B_p^{sq}(I\!R) \subset C(I\!R),$ *if* $s > \dfrac{1}{p}.$

Chapter 10

Statistical estimation using wavelets

10.1   Introduction

In Chapters 3, 5, 6 and 7 we discussed techniques to construct functions φ and ψ (father and mother wavelets), such that the wavelet expansion (3.5) holds for any function f in $L_2(I\!R)$. This expansion is a special kind of orthogonal series. It is "special", since unlike the usual Fourier series, the approximation is both in frequency and space. In this chapter we consider the problem of nonparametric statistical estimation of a function f in $L_2(I\!R)$ by wavelet methods. We study the density estimation and nonparametric regression settings. We also present empirical results of wavelet smoothing.

The idea of the estimation procedure is simple: we replace the unknown wavelet coefficients $\{\alpha_k\}, \{\beta_{jk}\}$ in the wavelet expansion (3.5) by estimates which are based on the observed data. This will require a truncation of the infinite series in (3.5) since we can only deal with a finite number of coefficients. In general, the truncation of the series and the replacement of wavelet coefficients in (3.5) will be done in a nonlinear way. We shall discuss in this chapter and in Chapter 11 how many basis functions we need and why a nonlinear procedure is necessary in order to automatically adapt to smoothness of the object being estimated.

Everywhere in this chapter we assume that the father and mother wavelets φ and ψ are real valued functions, rather than complex valued ones. This

125

covers the usual examples of Daubechies' wavelets, coiflets and symmlets.

The effect of nonlinear smoothing will become visible through many examples. We emphasize the fact that the statistical wavelet estimation technique may be of nonlinear form. The nonlinearity, introduced through thresholding of wavelet coefficients, guarantees smoothness adaptivity of the estimator as we shall see in Chapter 11.

10.2   Linear wavelet density estimation

Let $X_1, \ldots, X_n$ be independent identically distributed random variables with an unknown density f on $I\!\!R$. A straightforward wavelet estimator of f may be constructed by estimating the projection of f on V_{j_1} and it is defined as

$$\hat{f}_{j_1}(x) = \sum_k \hat{\alpha}_{j_0 k} \varphi_{j_0 k}(x) + \sum_{j=j_0}^{j_1} \sum_k \hat{\beta}_{jk} \psi_{jk}(x) \tag{10.1}$$

where $j_0, j_1 \in Z\!\!\!Z$ are some integers, and the values

$$\hat{\alpha}_{jk} = \frac{1}{n} \sum_{i=1}^n \varphi_{jk}(X_i), \tag{10.2}$$

$$\hat{\beta}_{jk} = \frac{1}{n} \sum_{i=1}^n \psi_{jk}(X_i) \tag{10.3}$$

are empirical estimates of the coefficients α_{jk} and β_{jk}, constructed by the method of moments. Note that $E(\hat{\alpha}_{jk}) = \alpha_{jk}$, $E(\hat{\beta}_{jk}) = \beta_{jk}$ (here and later $E(\cdot)$ denotes the expectation with respect to the joint distribution of observations), i.e. $\hat{\alpha}_{jk}$ and $\hat{\beta}_{jk}$ are unbiased estimators of α_{jk} and β_{jk}.

We assume below that φ and ψ are compactly supported. Remark that Proposition 8.6 (vi) yields in this case

$$\sum_k \varphi_{jk}(X_i)\varphi_{jk}(x) + \sum_k \psi_{jk}(X_i)\psi_{jk}(x) = \sum_k \varphi_{j+1,k}(X_i)\varphi_{j+1,k}(x) = K_{j+1}(x, X_i)$$

for any j, where the orthogonal projection kernels are

$$K_j(x, y) = 2^j K(2^j x, 2^j y), \quad K(x, y) = \sum_k \varphi(x - k)\varphi(y - k) \tag{10.4}$$

(as defined in Sections 8.3 and 8.5).

By successive application of this formula in (10.1), for j starting from j_0 up to j_1, we obtain:

$$\hat{f}_{j_1}(x) = \sum_k \hat{\alpha}_{j_1+1,k}\varphi_{j_1+1,k}(x) = \frac{1}{n}\sum_{i=1}^{n} K_{j_1+1}(x, X_i). \qquad (10.5)$$

The estimator $\hat{f}_{j_1}(x)$ is called *linear wavelet density estimator*. It is a linear function of the empirical measure

$$\nu_n = \frac{1}{n}\sum_{i=1}^{n}\delta_{\{x_i\}}$$

where $\delta_{\{x\}}$ is the Dirac mass at point x. Thus, $\hat{\alpha}_{jk} = \int \varphi_{jk}d\nu_n$, $\hat{\beta}_{jk} = \int \psi_{jk}d\nu_n$, and (10.1) may be formally viewed as a "wavelet expansion" for ν_n.

Unlike (3.5), where the expansion starts from $j = 0$, in (10.1) we have a series, starting from $j = j_0$ (the value j_0 may be negative, for example). This does not contradict the general theory, since nothing changes in the argument of Chapters 3, 5, 6, 7, if one considers the indices j, starting from j_0 instead of 0. In previous chapters the choice $j_0 = 0$ was made just to simplify the notation. Most software implementations set $j_0 = 0$. However, in practice the scaling effect may require a different choice for j_0. An empirical method of selecting j_0 is discussed in Section 11.5.

The role of the constant j_1 is similar to that of a bandwidth in kernel density estimation. The functions φ_{jk}, ψ_{jk} may be regarded as certain scaled "kernels", and their scale is defined by the value j which, in case of the estimator (10.1), is allowed to be in the interval $[j_0, j_1]$.

For applications there is no problem with the infinite series over k in (10.1). In fact, one implements only compactly supported wavelet bases (Haar, Daubechies, symmlets, coiflets). For these bases the sums $\sum_k \hat{\alpha}_{j_0 k}\varphi_{j_0 k}(x)$ and $\sum_k \hat{\beta}_{jk}\psi_{jk}(x)$ contain only a finite number of terms. The set of indices k included in the sums depends on the current value x.

REMARK 10.1 If *supp* $\psi \subseteq [-A, A]$, the sum $\sum_k \hat{\beta}_{jk}\psi_{jk}$ only contains the indices k such that

$$2^j \min_i X_i - A \le k \le 2^j \max_i X_i + A.$$

Hence, there are at the most $2^j(\max_i X_i - \min_i X_i) + 2A$ nonzero wavelet coefficients at the level j. If also the density f of X_i is compactly supported, the number M_j of non-zero wavelet coefficients on level j is $O(2^j)$.

The choice of resolution level j_1 in the wavelet expansion is important. Let us study this issue in more detail. Suppose that we know the exact regularity of the density, e.g. we assume that it lies in the *Sobolev class* of functions defined as follows:

$$W(m, L) = \{f : ||f^{(m)}||_2 \leq L, f \text{ is a probability density}\},$$

where $m > 1$ is an integer and $L > 0$ is a given constant. The number m denotes as in Section 8.2 the regularity of f. In Chapter 8 we introduced the Sobolev space $W_2^m(I\!R)$, here we just add the bound L on the L_2 norm of the derivative in an explicit form.

Let us investigate the behavior of the estimator defined in (10.1) when $f \in W(m, L)$. We consider its quadratic risk.

The mean integrated squared error (*MISE*) of any estimator $\hat{f}$ is

$$E||\hat{f} - f||_2^2 = E||\hat{f} - E(\hat{f})||_2^2 + ||E(\hat{f}) - f||_2^2.$$

This decomposition divides the risk into two terms:

- a *stochastic error* $E||\hat{f} - E(\hat{f})||_2^2$ due to the randomness of the observations.

- a *bias error* $||E(\hat{f}) - f||_2^2$ due to the method. This is the deterministic error made in approximating f by $E(\hat{f})$.

A fundamental phenomenon, common to all smoothing methods, appears in this situation. In fact, as it will be shown later, the two kinds of errors have antagonistic behavior when j_1 increases. The balance between the two errors yields an optimal j_1. Let us evaluate separately the bias and the stochastic error.

Bound for the bias error

In order to bound the bias term we shall draw upon results of Chapter 8. Recall some notation of Section 8.3 where approximation kernels were defined. According to this notation, the kernel $K(x, y)$ satisfies the Conditions

$H(N+1)$ and $M(N)$ for an integer $N > 0$, if for some integrable function $F(\cdot)$

$$|K(x,y)| \leq F(x-y), \text{ with } \int |x|^{N+1} F(x)dx < \infty, (\text{Condition} H(N+1)),$$

$$\int (y-x)^k K(x,y)dy = \delta_{0k}, \forall k = 0,1,\ldots,N, \forall x \in I\!R, (\text{Condition} M(N)).$$
$$(10.6)$$

We shall now apply the results of Chapter 8 for $m \leq N+1$. In the following it is assumed that φ satisfies Condition (θ) and $K(x,y)$ is the orthogonal projection kernel associated with φ (see Definition 8.7). The estimation of the bias error is merely a corollary of Theorem 8.1 (ii) and of the fact that

$$E\left(\hat{f}_{j_1}(x)\right) = E\left(K_{j_1+1}(x,X_1)\right) = K_{j_1+1}f(x) \tag{10.7}$$

(see (10.4) – (10.5) and the notation K_j in Section 8.3).

COROLLARY 10.1 *Suppose that the father wavelet φ is such that the pro-jection kernel*

$$K(x,y) = \sum_k \varphi(x-k)\varphi(y-k)$$

satisfies the condition (10.6). Then, for any $m \leq N+1$, there exists a constant $C > 0$ such that

$$\sup_{f \in W(m,L)} ||E(\hat{f}_{j_1}) - f||_2 \leq C 2^{-j_1 m}.$$

Bound for the stochastic error

PROPOSITION 10.1 *Suppose that φ is such that the kernel*

$$K(x,y) = \sum_k \varphi(x-k)\varphi(y-k)$$

satisfies $|K(x,y)| \leq F(x-y)$ with $F \in L_2(I\!R)$. Then we have

$$E||\hat{f}_{j_1} - E(\hat{f}_{j_1})||_2^2 \leq \frac{2^{j_1+1}}{n} \int F^2(v)dv.$$

Proof Using (10.7) we have

$$
\begin{aligned}
E\|\hat{f}_{j_1} - E(\hat{f}_{j_1})\|_2^2 &= E\int |\hat{f}_{j_1}(x) - E\{\hat{f}_{j_1}(x)\}|^2 dx \\
&= \int E\left(\frac{1}{n}\sum_{i=1}^n Y_i(x)\right)^2 dx,
\end{aligned}
$$

where $Y_i(x) = K_{j_1+1}(x, X_i) - E\left(K_{j_1+1}(x, X_i)\right)$ are i.i.d. zero-mean random variables.

Note that

$$
E\left(Y_i^2(x)\right) \leq E\left(K_{j_1+1}^2(x, X_i)\right) \leq 2^{2j_1+2}\int F^2\left(2^{j_1+1}(x - y)\right) f(y)dy. \quad (10.8)
$$

Thus

$$
\begin{aligned}
E\|\hat{f}_{j_1} - E(\hat{f}_{j_1})\|_2^2 &\leq \frac{2^{2j_1+2}}{n}\int \left[\int F^2\left(2^{j_1+1}(x - y)\right) dx\right] f(y)dy \\
&= \frac{2^{j_1+1}}{n}\int F^2(v)dv.
\end{aligned}
$$

We have used the Fubini theorem in the first inequality and a change of variable in the last equality. □

Later we write $a_n \simeq b_n$ for two positive sequences $\{a_n\}$ and $\{b_n\}$ if there exist $0 < A < B < \infty$ such that $A \leq \frac{a_n}{b_n} \leq B$ for n large enough.

The two bounds of Corollary 10.1 and Proposition 10.1 can be summarized in the following

THEOREM 10.1 *Under the assumptions of Proposition 10.1 and Corollary 10.1 we have that the MISE is uniformly bounded:*

$$
\sup_{f\in W(m,L)} E\|\hat{f}_{j_1} - f\|_2^2 \leq C_1\frac{2^{j_1}}{n} + C_2 2^{-2j_1 m},
$$

where C_1 and C_2 are positive constants. The RHS expression has a minimum when the two antagonistic quantities are balanced, i.e. for $j_1 = j_1(n)$ such that

$$
2^{j_1(n)} \simeq n^{\frac{1}{2m+1}}.
$$

In that case we obtain

$$
\sup_{f\in W(m,L)} E\|\hat{f}_{j_1(n)} - f\|_2^2 \leq Cn^{-\frac{2m}{2m+1}}, \quad (10.9)
$$

for some $C > 0$.

The result of Theorem 10.1 is quite similar to classical results on the L_2-convergence of the Fourier series estimates (see e.g. Centsov (1962), Pinsker (1980)). What is more interesting, wavelet estimators have good asymptotic properties not only in L_2, but also in general L_p norms, and not only on the Sobolev class $W(m, L)$, but also on functional classes defined by Besov constraints.

Here we give an example of such type of result. The following theorem is a generalization of Corollary 10.1, with the L_2 norm replaced by an L_p norm and the class $W(m, L)$ replaced by

$$B(s, p, q, L) = \{f : ||f||_{spq} \le L, f \text{ is a probability density}\}$$

where the norm $||f||_{spq}$ is the Besov norm defined in Section 9.2, and L is a finite constant. In the following we call $B(s, p, q, L)$ the *Besov class* of functions. It is the set of densities in a ball of radius L in the Besov space $B_p^{sq}(\mathbb{R})$.

THEOREM 10.2 *(Kerkyacharian & Picard (1992)). If*

$$K(x, y) = \sum_k \varphi(x - k)\varphi(y - k)$$

satisfies the conditions (10.6) with $F \in L_p(\mathbb{R})$, $0 < s < N + 1$, $2 \le p < \infty$, $1 \le q < \infty$, *then*

$$\sup_{f \in B(s,p,q,L)} E||\hat{f}_{j_1} - f||_p^p < C \left\{ 2^{-j_1 sp} + \left(\frac{2^{j_1}}{n}\right)^{p/2} \right\},$$

for some constant $C > 0$, *whenever* $2^{j_1} \le n$. *The RHS expression has a minimum when the two antagonistic terms are balanced, i.e. for* $j_1 = j_1(n)$ *such that*

$$2^{j_1(n)} \simeq n^{\frac{1}{2s+1}}.$$

In this case we obtain

$$\sup_{f \in B(s,p,q,L)} E||\hat{f}_{j_1(n)} - f||_p^p \le Cn^{-\frac{sp}{2s+1}},$$

for some $C > 0$.

REMARK 10.2 This bound is still true for $1 < p < 2$ if one requires in addition that $f(x) < w(x)$, $\forall x \in I\!R$, for some function $w \in L_{p/2}(I\!R)$ which is symmetric about a point $a \in I\!R^1$ and non-decreasing for $x > a$. One remarkable fact is that the level $j_1 = j_1(n)$ minimizing the bound of the risk still satisfies $2^{j_1(n)} \simeq n^{\frac{1}{2s+1}}$. Hence this choice is robust against variations of p, although it depends on the regularity s.

Proof of Theorem 10.2 is a slight modification of the above proofs for the L_2 case. We also split the risk into a stochastic term and a bias term:

$$E\|\hat{f}_{j_1} - f\|_p^p \leq 2^{p-1} \left\{ E\|\hat{f}_{j_1} - E(\hat{f}_{j_1})\|_p^p + \|E(\hat{f}_{j_1}) - f\|_p^p \right\}.$$

The bias term is treated similarly to Corollary 10.1, but using the approximation result of Theorem 9.5. The stochastic term requires in addition a moment inequality. In fact,

$$
\begin{aligned}
E\|\hat{f}_{j_1} - E(\hat{f}_{j_1})\|_p^p &= E \int |\frac{1}{n} \sum_{i=1}^{n} K_{j_1+1}(x, X_i) - E\{K_{j_1+1}(x, X_i)\}|^p dx \\
&= \int E \left(\left| \frac{1}{n} \sum_{i=1}^{n} Y_i(x) \right|^p \right) dx
\end{aligned}
$$

where $Y_i(x) = K_{j_1+1}(x, X_i) - E\{K_{j_1+1}(x, X_i)\}$ are i.i.d. centered random variables. Note also that $Y_i(x)$ are uniformly bounded by $2^{j_1+2}\|\theta_\varphi\|_\infty^2 < \infty$. In fact, Condition (θ) implies that $|K(x, y)| \leq \|\theta_\varphi\|_\infty^2$ (see Section 8.5). Thus, $|K_{j_1+1}(x, y)| \leq 2^{j_1+1}\|\theta_\varphi\|_\infty^2$.

The following proposition is proved in Appendix C.

PROPOSITION 10.2 (Rosenthal's inequality) *Let $p \geq 2$ and let $X_1, \ldots, X$ be independent random variables such that $E(X_i) = 0$ and $E(|X_i|^p) < \infty$. Then there exists $C(p) > 0$ such that*

$$E\left(\left| \sum_{i=1}^{n} X_i \right|^p \right) \leq C(p) \left\{ \sum_{i=1}^{n} E\left(|X_i|^p\right) + \left(\sum_{i=1}^{n} E(X_i^2) \right)^{p/2} \right\}.$$

COROLLARY 10.2 *If X_i are independent random variables such that $E(X_i) = 0$ and $|X_i| \leq M$, then for any $p \geq 2$ there exists $C(p) > 0$ such that:*

$$E\left(\left| \sum_{i=1}^{n} X_i \right|^p \right) \leq C(p) \left\{ M^{p-2} \sum_{i=1}^{n} E(X_i^2) + \left(\sum_{i=1}^{n} E(X_i^2) \right)^{p/2} \right\}.$$

Using this Corollary, we have

$$\frac{1}{n^p} E\left(\left|\sum_{i=1}^{n} Y_i(x)\right|^p\right) \le \frac{C(p)}{n^p} \left\{(2^{j_1+2}||\theta_\varphi||_\infty^2)^{p-2} \sum_{i=1}^{n} E\left(Y_i^2(x)\right) + \left(\sum_{i=1}^{n} E(Y_i^2(x))\right)^{p/2}\right\}.$$

As in the proof of Proposition 10.1, we find

$$\int \sum_{i=1}^{n} E(Y_i^2(x))dx \le n2^{j_1+1} \int F^2(v)dv.$$

It follows that

$$
\begin{aligned}
\int \frac{1}{n^p} E\left(\left|\sum_{i=1}^{n} Y_i(x)\right|^p\right) dx \quad &\le \quad C(p)(2||\theta_\varphi||_\infty^2)^{p-2} \frac{2^{(j_1+1)(p-1)}}{n^{p-1}} \int F^2(v)dv \\
&+ \quad C(p) \left(\frac{2^{j_1+1}}{n}\right)^{p/2} 2^{(j_1+1)p/2} \\
&\qquad \int \left[\int F^2\left(2^{j_1+1}(x-y)\right) f(y)dy\right]^{p/2} dx \\
&\le \quad C(p) \int F^2(v)dv(2||\theta_\varphi||_\infty^2)^{p-2} \left(\frac{2^{j_1+1}}{n}\right)^{p-1} \\
&+ \quad C(p) \int F^p(v)dv \left(\frac{2^{j_1+1}}{n}\right)^{p/2},
\end{aligned}
$$

where we used (10.8), Jensen's inequality and Fubini Theorem. To get the result of Theorem 10.2 it remains to observe that the leading term here is $\left(\frac{2^{j_1}}{n}\right)^{p/2}$ since $2^{j_1} \le n$ and $p \ge 2$ imply $\left(\frac{2^{j_1}}{n}\right)^{p-1} \le \left(\frac{2^{j_1}}{n}\right)^{p/2}$. □

Theorems 10.1 and 10.2 reflect the fact that, as a function of j_1, the bias decreases and the variance increases. In practice this means that with increasing level the linear wavelet estimates become rougher. This behavior can be seen from the following graphs.

In Figure 10.1 we show a graph with a uniform mixture probability density function and a wavelet estimate based on Haar basis wavelets with $j_0 = 0$ and $j_1 = 1$. The $n = 500$ pseudo random numbers are displayed as circles on the horizontal axis. One sees that the estimate at this resolution level is unable to capture the two peaks. We have chosen deliberately a uniform mixture density for this and the following examples. The power of wavelet local smoothing will become evident and the effects of different levels can be nicely demonstrated. The true density function has the form

$$f(x) = 0.5I\{x \in [0,1]\} + 0.3I\{x \in [0.4, 0.5]\} + 0.2I\{x \in [0.6, 0.8]\}$$

For practical wavelet density estimation, as well as in all simulated examples below, we use the technique slightly different from the original definition (10.1). An additional binning of data is introduced. The reason for this is to enable the use of discrete wavelet transform to compute the estimators (see Chapter 12). The binned density estimator is defined in $m = 2^K$ equidistant gridpoints $z_1, \ldots, z_m$, where $K \geq j_1$ is an integer, $z_l - z_{l-1} = \Delta > 0$. The computation is done in two steps. On the first step, using the data $X_1, \ldots, X_n$, one constructs a histogram, with bins of width Δ, centered at z_l. Usually this should be a very fine histogram, i.e. Δ should be relatively small. Let $\hat{y}_1, \ldots, \hat{y}_m$ be values of this histogram at points $z_1, \ldots, z_m$. On the second step one computes a certain approximation to the values

$$f_l = \sum_k \bar{\alpha}_{j_0 k} \varphi_{j_0 k}(z_l) + \sum_{j=j_0}^{j_1} \sum_k \bar{\beta}_{jk} \psi_{jk}(z_l), \ l = 1, \ldots, m, \qquad (10.10)$$

where

$$\bar{\alpha}_{jk} = \frac{1}{m} \sum_{i=1}^m \hat{y}_i \varphi_{jk}(z_i), \qquad (10.11)$$

$$\bar{\beta}_{jk} = \frac{1}{m} \sum_{i=1}^m \hat{y}_i \psi_{jk}(z_i). \qquad (10.12)$$

The approximately computed values f_l are taken as estimators of $f(z_l)$, $l = 1, \ldots, m$, at gridpoints $z_1, \ldots, z_m$. For more details on the computational algorithm and the effect of binning see Chapter 12. In the simulated example considered here we put $m = 256$.

The performance of an estimate $\hat{f}$ is expressed in terms of the integrated squared error

$$ISE = \int (\hat{f} - f)^2.$$

In our example we approximate the ISE as the squared difference of the density and its estimate at $m = 256$ gridpoints:

$$ISE \approx \frac{1}{m} \sum_{l=1}^m (f_l - f(z_l))^2.$$

The integrated squared error of $\hat{f} = \hat{f}_{j_1}$ with $j_1 = 1$ and $j_0 = 0$ is $ISE = 0.856$ which will be compared later with a kernel density estimate. Let us study now the effect of changing the level j_1. (From now on we shall set $j_0 = 0$.)

We first increase j_1 to 2. The corresponding estimate is given in Figure 10.2. As expected the estimate adapts more to the data and tries to resolve more local structure. The wavelet density estimate starts to model the peaks with a reduced ISE of 0.661.

This effect becomes more pronounced when we increase the level to $j_1 = 4$. The corresponding wavelet density estimate is shown in Figure 10.3. One sees that even more structure occurs and that the gap is modelled with the corresponding shoulders.

If we increase j_1 further the estimator becomes spiky. This can be seen from Figure 10.4 where we set $j_1 = 6$. Finally, for $j_1 = 8$ (i.e. $j_1 = \log_2 m$) the estimator reproduces the binned values $\hat{y}_1, \ldots, \hat{y}_m$ at gridpoints, (see Chapter 12 for more details) and this case is of no interest. Also, increasing j_1 above the value $\log_2 m$ makes no sense.

The ISE values for different wavelet bases are displayed in Table 10.1. The ISE values show as a function of j_1 the same overall behavior for all basis functions. The ISE values lie close together and the global minimum is achieved for j_1 around 4.

j_1	1	2	3	4	5	6	7
ISE(D2)	0.857	0.661	0.290	0.224	0.141	0.191	0.322
ISE(D4)	0.747	0.498	0.269	0.156	0.125	0.190	0.279
ISE(D8)	0.698	0.650	0.459	0.147	0.128	0.158	0.260
ISE(D16)	0.634	0.613	0.465	0.132	0.133	0.186	0.296
ISE(S4)	0.700	0.539	0.319	0.146	0.104	0.142	0.275
ISE(S8)	0.625	0.574	0.328	0.140	0.135	0.147	0.310
ISE(C1)	0.595	0.558	0.503	0.168	0.136	0.170	0.306

Table 10.1: ISE values for different density estimates

Summarizing this experiment of changing the level j_1 we find an illustration to the effect given in Corollary 10.1 and Proposition 10.1. The parameter j_1 determines the spikyness or frequency localization of the estimate. The

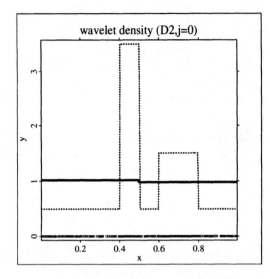

Figure 10.1: Uniform mixture random variables $(n = 500)$ with density and a Haar wavelet estimate with $j_1 = 1$.

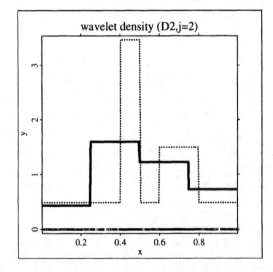

Figure 10.2: The same variables as in Figure 10.1 and a Haar wavelet estimate with $j_1 = 2$.

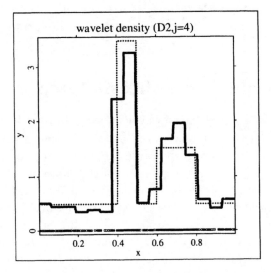

Figure 10.3: The same variables as in Figure 10.1 and a Haar wavelet density estimate with $j_1 = 4$.

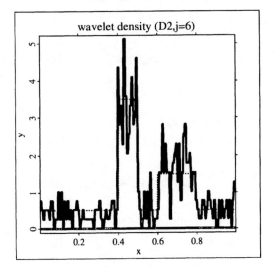

Figure 10.4: Haar wavelet density estimate with $j_1 = 6$.

more levels we let into (10.1) the more spiky the estimate becomes. The bias decreases but the variance increases, and there is an optimum at j_1 around 4.

10.3   Soft and hard thresholding

Figure 10.4 shows that the linear wavelet estimator may have small spikes. This reflects the fact that unnecessary high oscillations are included. Since the detail coefficients β_{jk} are responsible for such oscillations, it is thus natural to introduce a selection procedure for β_{jk}'s. More precisely we suppress too small coefficients by introduction of a threshold. Such a procedure is called wavelet thresholding. There exist various thresholding procedures. Here we introduce two of them: soft thresholding and hard thresholding. These techniques were proposed by D.Donoho and I.Johnstone in the beginning of 1990-ies. A more detailed survey of wavelet thresholding methods is deferred to Chapter 11.

In *soft thresholding* one replaces $\hat{\beta}_{jk}$ in (10.1) by

$$\hat{\beta}_{jk}^S = (|\hat{\beta}_{jk}| - t)_+ \, \text{sign} \, (\hat{\beta}_{jk}) \qquad (10.13)$$

where $t > 0$ is a certain threshold. The wavelet estimator with soft thresholding is also called *wavelet shrinkage* estimator since it is related to Stein's shrinkage (see Section 11.5). In *hard thresholding* one replaces $\hat{\beta}_{jk}$ in (10.1) by

$$\hat{\beta}_{jk}^H = \hat{\beta}_{jk} \, I\{|\hat{\beta}_{jk}| > t\}. \qquad (10.14)$$

The plots of $\hat{\beta}_{jk}^S, \hat{\beta}_{jk}^H$ versus $\hat{\beta}_{jk}$ are shown in Figure 10.5.

The *wavelet thresholding density estimator* is defined as:

$$f_n^*(x) = \sum_k \hat{\alpha}_{j_0 k} \varphi_{j_0 k}(x) + \sum_{j=j_0}^{j_1} \sum_k \beta_{jk}^* \psi_{jk}(x), \qquad (10.15)$$

where $\beta_{jk}^* = \hat{\beta}_{jk}^S$ (soft thresholding) or $\beta_{jk}^* = \hat{\beta}_{jk}^H$ (hard thresholding).

The effect of thresholding is shown in Figures 10.6 – 10.11 for the same sample as in the previous graphs. Figure 10.6 shows the wavelet density estimator ($j_1 = 8$, Haar $D2$) with hard threshold value t set to $0.4 \max\limits_{j,k} |\hat{\beta}_{jk}|$. We see that spikes are present. This effect is less pronounced if we increase

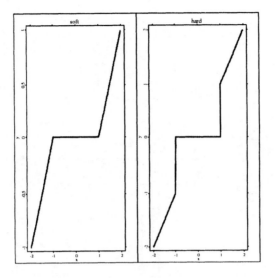

Figure 10.5: Soft and hard thresholding.

the threshold to $0.6 \max_{j,k} |\hat{\beta}_{jk}|$, see Figure 10.7.

We increase the threshold value further to $0.8 \max_{j,k} |\hat{\beta}_{jk}|$ so that only two coefficients are passing the threshold, see Figure 10.8. We see that increasing the threshold value produces smoother wavelet density estimates but still has visible local variation. This effect is avoided by soft thresholding.

The soft threshold was set equal to $0.8 \max_{j,k} |\hat{\beta}_{jk}|$ for Figure 10.9. The following Figure 10.10 shows the estimate with a soft threshold of $0.6 \max_{j,k} |\hat{\beta}_{jk}|$. In comparison with Figure 10.7 one sees the effect of downweighting the coefficients. Figure 10.11 finally shows the threshold value decreased to $0.4 \max_{j,k} |\hat{\beta}_{jk}|$. The estimate is rougher due to the lower threshold value.

In our specific example soft thresholding decreased the ISE further. In Table 10.2 we give estimates of the integrated squared error distances $ISE(f_n^*, f) = \int (f_n^* - f)^2$ as a function of the threshold value and the method of hard or soft thresholding. One sees that the best ISE value is obtained for soft thresholding procedure with $j_1 = 8, t = 0.4 \max_{j,k} |\hat{\beta}_{j,k}|$. However, this is not the best case, if one compares Figures 10.6 – 10.11 visually. The L_2 error

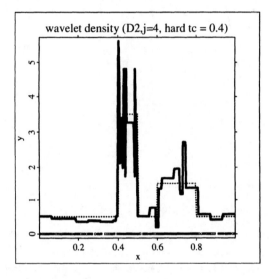

Figure 10.6: A sample of $n = 500$ points with uniform mixture density and a Haar wavelet density estimate. The hard threshold was set to $0.4 \max_{j,k} |\hat{\beta}_{jk}|$.

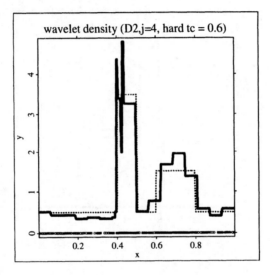

Figure 10.7: A sample of $n = 500$ points with density and a Haar wavelet density estimate. The hard threshold was set to $0.6 \max_{j,k} |\hat{\beta}_{jk}|$.

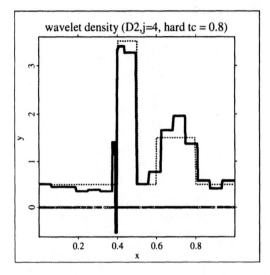

Figure 10.8: A sample of $n = 500$ points with density and a Haar wavelet density estimate. The hard threshold was set to $0.8 \max_{j,k} |\hat{\beta}_{jk}|$.

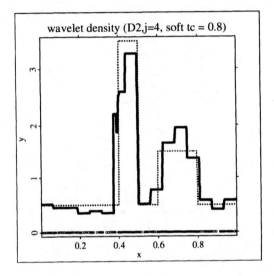

Figure 10.9: Soft thresholding with data from Figure 10.6. Threshold value $0.8 \max_{j,k} |\hat{\beta}_{jk}|$.

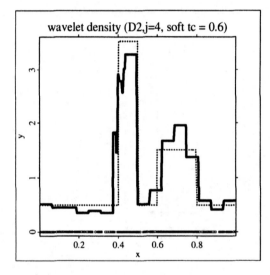

Figure 10.10: Soft thresholding with data from Figure 10.6. Threshold value $0.6 \max_{j,k} |\hat{\beta}_{jk}|$.

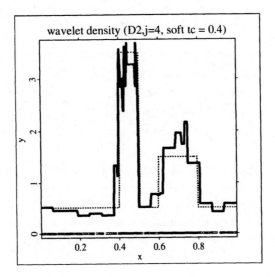

Figure 10.11: Soft thresholding with data from Figure 10.6. Threshold value $0.4 \max_{j,k} |\hat{\beta}_{jk}|$.

(*ISE* or *MISE*) is not always adequate for visual interpretation (cf. Marron & Tsybakov (1995)).

threshold/$\max_{j,k} |\hat{\beta}_{j,k}|$	0.4	0.6	0.8
hard	0.225	0.193	0.201
soft	0.177	0.221	0.253

Table 10.2: *ISE* for different threshold values, $j_1 = 8$, Haar wavelet.

Remark that we choose thresholds as multiples of $\max_{j,k} |\hat{\beta}_{j,k}|$, in order to compare them on a common scale. Thresholding can be done level by level, allowing $t = t_j$ to depend on the level j. Then the values t_j can be chosen as multiples of $\max_{k} |\hat{\beta}_{j,k}|$. Another natural way of choosing a threshold is taking t or t_j as an order statistic of the set of absolute values of coefficients $\{|\hat{\beta}_{j,k}|\}_{j,k}$ or $\{|\hat{\beta}_{j,k}|\}_k$, respectively. This is discussed in Section 11.5.

As a further reference to later chapters we give a modification of the above figures that avoids the local spikyness visible in the last graphs. Figure 10.12 presents a so called translation invariant wavelet density smoother. To construct it we essentially perform an average of as many wavelet smoothers as there are bins. In Section 12.5 we define this estimator.

10.4   Linear versus nonlinear wavelet density estimation

In Section 10.2 we studied the linear wavelet methods. The word *linear* is referring to the fact that the estimator is a linear function of the empirical measure $\nu_n = \frac{1}{n} \sum_{i=1}^{n} \delta_{\{X_i\}}$ ($\delta_{\{x\}}$ is the Dirac mass at point x). Then we have seen in Section 10.3 a need for a (non-linear) thresholding type selection procedure on the coefficients β_{jk} coming from a practical point of view. This suggests that for practical reasons non-linear estimators may be useful. We are going to show now that there is also a theoretical need for non-linear estimators.

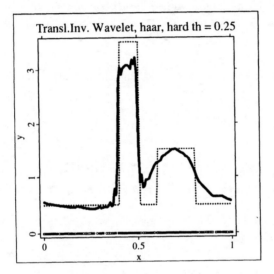

Figure 10.12: Translation invariant thresholding with data from Figure 10.6. Threshold value $0.25 \max_{j,k} |\hat{\beta}_{jk}|$.

Note that the linear procedures of Section 10.2 are robust with respect to the parameters p and q of Besov classes in the sense that the best choice of the level $j_1(n)$ depends only on the regularity s (cf. Remark 10.2). Observe also that in Theorem 10.2 the function f belongs to the class $B(s, p, q, L)$, and the risk of an estimator is calculated in L_p norm, with the same p as in the definition of the class. This will be referred to as *matched a priori assumptions* on the smoothness class of functions f and on the risk.

The following questions arise then:

Question 10.1 *What is the optimal rate of convergence attainable by an estimator when the underlying function f belongs to a certain Besov class of functions ?*

Question 10.2 *Is there an effect of matched a priori assumptions in this optimal rate ?*

Question 10.3 *Does it happen that linear wavelet estimators attain the optimal rate of convergence ?*

Question 10.4 *If this is the case, is it always true or are there situations where one must use non-linear procedures to obtain optimal rates ?*

Question 10.5 *If it is the case, what about the performance of wavelet thresholding estimators?*

The aim of this section is to answer these questions. To define correctly the notion of optimal rate of convergence, let us introduce the following minimax framework.

Let V be a class of functions. Assume that it is known that $f \in V$. The L_p *risk* of an arbitrary estimator $T_n = T_n(X_1, ..., X_n)$ based on the sample $X_1, .., X_n$ is defined as

$$E||T_n - f||_p^p,$$

$1 \leq p < \infty$.

Consider the L_p *minimax risk*:

$$R_n(V, p) = \inf_{T_n} \sup_{f \in V} E||T_n - f||_p^p,$$

where the infimum is taken over all estimators T_n (measurable functions taking their values in a space containing V) of f. Let us also consider the *linear L_p minimax risk*

$$R_n^{lin}(V,p) = \inf_{T_n^{lin}} \sup_{f \in V} E\|T_n^{lin} - f\|_p^p$$

where the infimum is now taken over all linear estimators T_n^{lin} in the sense quoted above. Obviously,

$$R_n^{lin}(V,p) \geq R_n(V,p). \tag{10.16}$$

DEFINITION 10.1 *The sequence $a_n \simeq R_n(V,p)^{\frac{1}{p}}$ is called* **optimal rate of convergence** *(or minimax rate of convergence) on the class V for the L_p risk. We say that an estimator f_n of f attains optimal rate of convergence if*

$$\sup_{f \in V} E\|f_n - f\|_p^p \simeq R_n(V,p).$$

Note that the *optimal rate of convergence* is defined up to a constant or bounded variable factor.

In view of this definition, the answer to Question 10.1 would be obtained by investigation of the asymptotics of the minimax risk $R_n(V,p)$, when V is a Besov class. Note that some information on this asymptotics is already available from Theorem 10.2. In fact, Theorem 10.2 implies that if $V = B(s,p,q,L)$, then

$$R_n^{lin}(V,p) \leq Cn^{-\frac{sp}{2s+1}}, \tag{10.17}$$

where $C > 0$ is a constant.(Here and later we use generic notation C for positive constants, possibly different.)

If, in addition, we could prove that, for $V = B(s,p,q,L)$ and some $C' > 0$

$$R_n(V,p) \geq C'n^{-\frac{sp}{2s+1}}, \tag{10.18}$$

then it would follow from (10.16) and (10.17) that

$$R_n^{lin}(V,p) \simeq R_n(V,p) \simeq n^{-\frac{sp}{2s+1}} \tag{10.19}$$

and the linear estimators introduced in Section 10.2 would attain the optimal rate which would be $n^{-\frac{s}{2s+1}}$. This would give an answer to Questions 10.1

and 10.2. However, Theorem 10.2, that we used in this reasoning, was proved only for the *matched* case. In the non-matched case, where $V = B(s, r, q, L)$ and $r \neq p$, the situation turns out to be more complex. The minimax rates of convergence are, in general, different from $n^{-\frac{s}{2s+1}}$, and they depend on the configuration (s, r, p, q). Moreover, it is not always possible to achieve optimal rates by use of linear estimators. Before discussing this in more detail, let us make some remarks on related earlier work in minimax nonparametric estimation.

The minimax theory has been largely developed in 1980-ies and 1990-ies. A variety of results have been obtained with different function classes, losses and observation models. Among many others let us mention Bretagnolle & Huber (1979), Ibragimov & Hasminskii (1980, 1981), Stone (1980, 1982), Birgé (1983), who obtained, in particular, the minimax rates for Sobolev classes and L_p risks and proved that kernel estimators attain these rates under certain conditions. Pinsker (1980), Efroimovich & Pinsker (1981), Nussbaum (1985) obtained not only rate optimal but exact asymptotically optimal procedures for the L_2 risks on Sobolev classes. In all these results the risk function is *matched* with the class of functions.

The first systematic study of non-matched situation is due to Nemirovskii (1985). He classified optimal convergence rates (up to a logarithmic factor) for L_r Sobolev classes and L_p risks, in the nonparametric regression problem with regular design.

Nemirovskii, Polyak & Tsybakov(1983, 1985) and Nemirovskii (1986) pointed out that for certain combinations of L_p risks and Sobolev classes no linear estimator can attain optimal rates in nonparametric regression and the best nonlinear estimators outperform the linear ones by a factor polynomial in n. In other words, kernel, spline, Fourier or linear wavelet methods even though properly windowed are suboptimal. This is what we are going to investigate below in the case of density estimation, Besov classes and L_p risks.

As compared to Section 10.2, we use for technical reasons a slightly modified definition of Besov classes. We add the compactness of support assumption on the density f. Let $s > 0, r \geq 1, q \geq 1, L > 0, L' > 0$ be fixed numbers. Consider the *Besov class* $\tilde{B}(s, r, q, L, L') = \tilde{B}(s, r, q)$ defined as follows:

$$\tilde{B}(s, r, q) = \{f : f \text{ is a probability density on } I\!\!R \text{ with a compact support}$$
$$\text{of length} \leq L', \text{ and } ||f||_{srq} \leq L\}.$$

The entries L and L' are omitted in the notation for sake of brevity.

THEOREM 10.3 *Let* $1 \leq r \leq \infty$, $1 \leq q \leq \infty$, $s > \frac{1}{r}$, $1 \leq p < \infty$. *Then there exists* $C > 0$ *such that*

$$R_n(\tilde{B}(s,r,q),p) \geq Cr_n(s,r,p,q), \qquad (10.20)$$

where

$$r_n(s,r,p,q) = \begin{cases} n^{-\alpha_1 p}, & \alpha_1 = \frac{s}{2s+1}, & \text{if } r > \frac{p}{2s+1}, \\ \left(\frac{\log n}{n}\right)^{\alpha_2 p}, & \alpha_2 = \frac{s - \frac{1}{r} + \frac{1}{p}}{2(s - \frac{1}{r}) + 1}, & \text{if } r \leq \frac{p}{2s+1}. \end{cases} \qquad (10.21)$$

Let, moreover, $s' = s - \left(\frac{1}{r} - \frac{1}{p}\right)_+$. *Then*

$$R_n^{lin}(\tilde{B}(s,r,q),p) \simeq n^{-\frac{s'p}{2s'+1}}. \qquad (10.22)$$

This theorem has been proved in Donoho, Johnstone, Kerkyacharian & Picard (1996). (We refer to this paper later on, for further discussion.) Before the proof of Theorem 10.3 some remarks and a corollary are in order.

REMARK 10.3 The result (10.20) is a lower bound on the minimax risk over the Besov classes. It divides the whole space of values (r,p) into two main zones:

$$\text{(i) } r > \frac{p}{2s+1} \quad (regular\ zone),$$

and

$$\text{(ii) } r \leq \frac{p}{2s+1} \quad (sparse\ zone).$$

The names "regular" and "sparse" are motivated as follows.

The regular zone is characterized by the same rate of convergence $n^{-\frac{s}{2s+1}}$ as in the matched case. It will be clear from the proof of (10.20) that the worst functions f (i.e. the hardest functions to estimate) in the regular case are of a saw-tooth form: their oscillations are equally dispersed on a fixed interval of the real line.

The sparse zone is characterized by a different rate of convergence, as compared to the matched case. The hardest functions to estimate in this zone have quite sharply localized irregularities, and are very regular elsewhere.

Thus, only few detail coefficients β_{jk} are non-zero. This explains the name "sparse".

The boundary $r = \frac{p}{2s+1}$ between the sparse and regular zones is a special case. Here $\alpha_2 = \alpha_1$, and the rate r_n differs from that of the regular zone only by a logarithmic factor.

REMARK 10.4 The result (10.22) on linear risks also splits their asymptotics into two zones. In fact, s' takes two possible values:

$$s' = \begin{cases} s & , \quad \text{if} \quad r \geq p, \\ s - \frac{1}{r} + \frac{1}{p} & , \quad \text{if} \quad r < p. \end{cases}$$

Thus, we have the zones:

$$\text{(i) } r \geq p \text{ (}homogeneous\ zone\text{)},$$

and

$$\text{(ii) } r < p \text{ (}non\text{-}homogeneous\ zone\text{)}.$$

In the homogeneous zone linear estimators attain the rate of convergence $n^{-\frac{s}{2s+1}}$ of the matched case. In the non-homogeneous zone we have $s' = s - \frac{1}{r} + \frac{1}{p} < s$, and thus the convergence rate of linear estimators $n^{-\frac{s'}{2s'+1}}$ is slower than $n^{-\frac{s}{2s+1}}$.

Note that the homogeneous zone is always contained in the regular zone. Thus, we have the following corollary.

COROLLARY 10.3 (Homogeneous case) *Let $r \geq p$. Then, under the assumptions of Theorem 10.3,*

$$R_n^{lin}(\tilde{B}(s,r,q),p) \simeq R_n(\tilde{B}(s,r,q),p) \simeq n^{-\frac{sp}{2s+1}}.$$

Graphically, the Remarks 10.3 and 10.4 can be summarized as shown in Figure 10.13. (Intermediate zone is the intersection of regular and non-homogeneous zones.) The 3 zones in Figure 10.13 are characterized as follows:

- *homogeneous zone:*

    - optimal rate is $n^{-\frac{s}{2s+1}}$, as in the matched case,

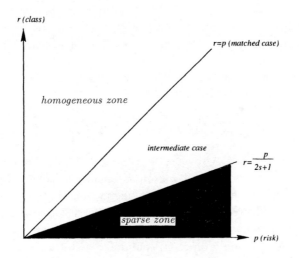

Figure 10.13: Classification of optimal rates of convergence for linear and non-linear estimates.

- linear estimators attain the optimal rate,

- *intermediate zone*:

  - optimal rate is $n^{-\frac{s}{2s+1}}$, as in the matched case,
  - linear estimators do not attain the optimal rate,

- *sparse zone*:

  - optimal rate is slower than in the matched case, and it depends on p and r,
  - linear estimators do not attain the optimal rate.

This classification contains answers to the Questions 10.2, 10.3 and 10.4. In doing this classification, we tacitly assumed that the values r_n in (10.21) represent not only the lower bounds for minimax risks, but also their true asymptotics. This assumption will be justified (to within logarithmic factors of the rates) in the next section.

The rest of this section is devoted to the proof of Theorem 10.3. We give the complete proof of (10.20), and some remarks on the proof of (10.22), referring for more details to Donoho, Johnstone, Kerkyacharian & Picard (1996).

Consider first the proof of (10.22). Since $\tilde{B}(s, p, q, L, L') \subset B(s, p, q, L)$, $\forall L' > 0$, it follows from Theorem 10.2 that

$$R_n^{lin}(\tilde{B}(s, p, q), p) \le Cn^{-\frac{sp}{2s+1}} \tag{10.23}$$

where $C > 0$ is a constant. On the other hand, consider the linear estimator $\hat{f}_{j_1}$ such that the functions φ and ψ are compactly supported and the conditions of Theorem 10.2 are satisfied. Then, using the fact that $f \in \tilde{B}(s, p, q)$ is compactly supported, we get that $\hat{f}_{j_1}$ has a support contained in a δ- neighborhood of $supp\ f$, where $\delta > 0$ depends only on φ, ψ and j_0. Thus, there exists $\overline{C} > 0$ depending only on φ, ψ, j_0 and L', such that $supp\ (\hat{f}_{j_1} - f)$ has a length $\le \overline{C}$. Using this and the Hölder inequality, we obtain, for $r > p$,

$$E\|\hat{f}_{j_1} - f\|_p^p \le \overline{C}^{1-p/r}(E\|\hat{f}_{j_1} - f\|_r^r)^{p/r}$$

and hence, in view of Theorem 10.2 with $2^{j_1} \simeq n^{\frac{1}{2s+1}}$,

$$R_n^{lin}(\tilde{B}(s, r, q), p) \le Cn^{-\frac{sp}{2s+1}}, r > p, \tag{10.24}$$

where $C > 0$ is a constant. For $r < p$ using the embedding theorems of Besov spaces (see Corollary 9.2), we have $\tilde{B}(s, r, q) \subset \tilde{B}(s', p, q)$ with $s' = s + \frac{1}{p} - \frac{1}{r}$ and so, in view of (10.23),

$$R_n^{lin}(\tilde{B}(s, r, q), p) \le R_n^{lin}(\tilde{B}(s', p, q), p) \le$$
$$\le Cn^{-\frac{s'p}{2s'+1}}, r < p. \tag{10.25}$$

Combining (10.23)–(10.25), we find

$$R_n^{lin}(\tilde{B}(s, r, q), p) \le Cn^{-\frac{s'p}{2s'+1}}, \tag{10.26}$$

for all (r, p) satisfying the assumptions of Theorem 10.3. Next clearly,

$$R_n^{lin}(\tilde{B}(s, r, q), p) \ge R_n(\tilde{B}(s, r, q), p),$$

which, together with (10.20) implies

$$R_n^{lin}(\tilde{B}(s,r,q),p) \geq C'n^{-\frac{sp}{2s+1}}, r \geq p, \qquad (10.27)$$

where $C' > 0$ is a constant.

From (10.26) and (10.27) we deduce (10.22) in the homogeneous case (i.e. for $r \geq p$). To show (10.22) in the case $r < p$ one needs to complete (10.26) by the lower bound

$$R_n^{lin}(\tilde{B}(s,r,q),p) \geq C'n^{-\frac{s'p}{2s'+1}}, r < p,$$

with some $C' > 0$. For the proof of this bound we refer to Donoho, Johnstone, Kerkyacharian & Picard (1996).

It remains to prove the lower bound (10.20). The proof presented below differs from that of Donoho, Johnstone, Kerkyacharian & Picard (1996). We employ different techniques for the sparse and regular cases respectively.

In the sparse case, we use a simple lemma, due to Korostelev & Tsybakov (1993b), Ch.2, which yields a lower bound in the problem of distinguishing between a finite number of hypotheses in terms of the behavior of the likelihood ratio. This technique is flexible enough to be implemented in a variety of situations (see e.g. Hoffmann (1996) for application to estimation of a volatility function in a stochastic differential equation). Further refinements of this lemma are given in Korostelev & Tsybakov (1993a) and Tsybakov (1995). For convenience we formulate this lemma here and give its proof.

In the regular case, the proof of (10.20) is based on Assouad's lemma (see Bretagnolle & Huber (1979), Assouad (1983), Korostelev & Tsybakov (1993b),Ch.2)).

We start with the proof of the lower bound (10.20) in the sparse case.

Risk bounds: sparse case

Let $d(\cdot,\cdot)$ be a distance on V and let

$$\Lambda_n(f,g) = \frac{dP_f^n}{dP_g^n}$$

be the likelihood ratio where P_f^n is the probability distribution of $X_1,\ldots,X_n$ if f is true. The ratio $\Lambda_n(f,g)$ is defined only if P_f^n is absolutely continuous with respect to P_g^n.

LEMMA 10.1 *(Korostelev & Tsybakov (1993b)) Let V contain the functions $g_0, \dots, g_K$ such that*

(i) $d(g_{k'}, g_k) \geq \delta > 0$, for $k = 0, \dots, K$, $k \neq k'$,

(ii) $K \geq \exp(\lambda_n)$, for some $\lambda_n > 0$,

(iii) $\Lambda_n(g_0, g_k) = \exp\{z_n^k - v_n^k\}$ where z_n^k is a random variable such that there exists $\pi_0 > 0$ with $P_{g_k}^n(z_n^k > 0) \geq \pi_0$, and v_n^k are constants,

(iv) $\sup_k v_n^k \leq \lambda_n$.

Then

$$\sup_{f \in V} P_f^n\left(d(\hat{f}, f) \geq \frac{\delta}{2}\right) \geq \sup_{1 \leq k \leq K} P_{g_k}^n(d(\hat{f}, g_k) \geq \frac{\delta}{2}) \geq \frac{\pi_0}{2},$$

for an arbitrary estimator $\hat{f}$.

Proof.   Let us observe that because of the triangle inequality $d(g_i, g_k) \leq d(\hat{f}, g_i) + d(\hat{f}, g_k)$, the events $\{d(\hat{f}, g_i) < \frac{\delta}{2}\}$ are disjoint and

$$
\begin{aligned}
P_{g_0}^n\left(d(\hat{f}, g_0) \geq \frac{\delta}{2}\right) &\geq P_{g_0}^n\left(\cup_{i \neq 0}\{d(\hat{f}, g_i) < \frac{\delta}{2}\}\right) \\
&= \sum_{i \neq 0} P_{g_0}^n\left(d(\hat{f}, g_i) < \frac{\delta}{2}\right) \\
&= \sum_{i \neq 0} E_{g_i}^n\left[\Lambda_n(g_0, g_i) I\left(d(\hat{f}, g_i) < \frac{\delta}{2}\right)\right] \\
&\geq \sum_{i \neq 0} \exp(-v_n^i) P_{g_i}^n\left(d(\hat{f}, g_i) < \frac{\delta}{2}, z_n^i > 0\right) \\
&\geq \exp(-\lambda_n) \sum_{i \neq 0} P_{g_i}^n\left(d(\hat{f}, g_i) < \frac{\delta}{2}, z_n^i > 0\right),
\end{aligned}
$$

where E_g^n denotes expectation with respect to P_g^n. Assume that

$$P_{g_i}^n\left(d(\hat{f}, g_i) \geq \frac{\delta}{2}\right) \leq \frac{\pi_0}{2}$$

for all $i \neq 0$ (if it is not the case, the lemma is proved).

Therefore, $P_{g_i}^n\left(d(\hat{f}, g_i) < \frac{\delta}{2}\right) \geq 1 - \frac{\pi_0}{2}$, and since $P_{g_i}^n(z_n^i > 0) \geq \pi_0$, then

$$P_{g_i}^n\left(d(\hat{f}, g_i) < \frac{\delta}{2}; z_n^i > 0\right) \geq \frac{\pi_0}{2},$$

for all $i \neq 0$. It follows that

$$P_{g_0}^n \left(d(\hat{f}, g_0) \geq \frac{\delta}{2} \right) \geq \frac{\pi_0}{2} K \exp(-\lambda_n) \geq \frac{\pi_0}{2}.$$

$\square$

Let us now use Lemma 10.1 to prove the lower bound on the minimax risk in the sparse case: $r \leq \frac{p}{2s+1}$. Consider a function g_0 such that

- g_0 is a probability density,

- $||g_0||_{srq} \leq \frac{L}{2}$,

- $g_0(x) = c_0 > 0$ on an interval $[a, b]$, $a < b$,

- the length of *supp* g_0 is less than L'.

Clearly $g_0 \in \tilde{B}(s, r, q)$. Let ψ be a very regular (for example satisfying the assumptions of Theorem 9.6) wavelet with compact support (see Chapter 7). Consider the set $\{g_k = g_0 + \gamma \psi_{jk}, k \in R_j\}$, where j is an integer to be chosen below ,$\gamma > 0$, and R_j is the maximal subset of $\mathbb{Z}$ such that

$$supp \; \psi_{jk} \subset [a, b], \qquad \forall k \in R_j,$$
$$supp \; \psi_{jk} \cap supp \; \psi_{jk'} = \emptyset, \quad if \; k \neq k'.$$

It is easy to see that g_k are probability densities. In fact $\int \psi_{jk} = 0$ as follows from (5.18). Note that card $R_j \simeq \frac{2^j}{T}(b - a)$ if T is the length of the support of ψ. Assume that T, ψ and a, b are chosen so that, for our value j,

$$S_j = \text{card } R_j = 2^j.$$

Using Corollary 9.1 we have $||g_k||_{srq} \leq ||g_0||_{srq} + \gamma ||\psi_{jk}||_{srq} \leq \frac{L}{2} + c_1 \gamma 2^{j(s+\frac{1}{2}-\frac{1}{r})}$, where $c_1 > 0$ is a constant ; in what follows we set $c_1 = 1$ for simplicity. Also $\int g_k = 1; g_k(x) \geq c_0 - \gamma ||\psi||_{\infty} 2^{\frac{j}{2}}, \forall x \in [a, b]$, and the length of *supp* g_k is less than L'. Hence $g_k \in \tilde{B}(s, r, q)$ if $\gamma \leq c_0 2^{-j/2}/||\psi||_{\infty}$ and $\gamma \leq \frac{L}{2} 2^{-j(s+\frac{1}{2}-\frac{1}{r})}$. Obviously, the first inequality is true for j large enough if the second inequality is satisfied. In the following we assume that this is the case.

We have for the L_p distance $d(\cdot, \cdot)$:

$$d(g_k, g_{k'}) \geq d(g_k, g_0) = ||g_k - g_0||_p = \gamma 2^{j(\frac{1}{2}-\frac{1}{p})} ||\psi||_p, \; k \neq 0, \; k' \neq 0, k \neq k'.$$

Thus, condition (i) of Lemma 10.1 holds with $\delta = \gamma 2^{j(\frac{1}{2}-\frac{1}{p})}||\psi||_p$. The measures $P_{g_0}^n$ and $P_{g_k}^n$ are mutually absolutely continuous with

$$
\begin{aligned}
\Lambda_n(g_0, g_k) &= \prod_{i=1}^n \frac{g_0(X_i)}{g_k(X_i)} \\
&= \prod_{i=1}^n \left(1 - \frac{\frac{\gamma}{c_0}\psi_{jk}(X_i)}{1 + \frac{\gamma}{c_0}\psi_{jk}(X_i)} \right) \\
&= \exp\left\{ \sum_{i=1}^n \left[\frac{\gamma}{c_0}V_k(X_i) - \frac{1}{2}\left(\frac{\gamma}{c_0}V_k(X_i)\right)^2 + \kappa\left(\frac{\gamma}{c_0}V_k(X_i)\right) \right] \right\},
\end{aligned}
$$

where we denote

$$
V_k(X_i) = \frac{\psi_{jk}(X_i)}{1 + \frac{\gamma}{c_0}\psi_{jk}(X_i)} = \frac{c_0 \psi_{jk}(X_i)}{g_k(X_i)}
$$

and $\kappa(u) = \log(1-u) - u + \frac{u^2}{2}$.

Now, choose

$$
\frac{\gamma}{c_0} = t_0 \sqrt{\frac{\log n}{n}}, \quad 2^j \simeq \left(\sqrt{\frac{n}{\log n}}\right)^{\frac{1}{s+\frac{1}{2}-\frac{1}{r}}}
$$

where $t_0 > 0$, and let us verify that we can apply Lemma 10.1. Put

$$
\begin{aligned}
v_n^k &= \frac{t_0^2}{2} E_{g_k}^n \{V_k(X_i)^2\} \log n, \\
\zeta_n &= t_0 \sqrt{\frac{\log n}{n}} \sum_{i=1}^n V_k(X_i), \\
\eta_n &= \sum_{i=1}^n \kappa\left(\frac{\gamma}{c_0}V_k(X_i)\right) - \frac{t_0^2 \log n}{2n} \sum_{i=1}^n \left(V_k(X_i)^2 - E_{g_k}^n\{V_k(X_i)^2\}\right).
\end{aligned}
$$

We have $\Lambda_n(g_0, g_k) = \exp\{z_n^k - v_n^k\}$ with $z_n^k = \zeta_n + \eta_n$. (We omitted the index k in ζ_n or η_n). Now, let us observe that $s > 1/r$ and thus for j large enough we have $g_k(u) > \frac{c_0}{2}$, $\forall u \in [a, b]$. Hence,

$$
E_{g_k}^n\{V_k(X_i)^2\} \leq 2c_0 \int \psi_{jk}^2(u)du = 2c_0, \tag{10.28}
$$

$$
E_{g_k}^n\{|V_k(X_i)|^3\} \leq 4c_0 \int |\psi_{jk}(u)|^3 du \leq 2^{j/2+2}||\psi||_3^3 c_0, \tag{10.29}
$$

$$
E_{g_k}^n\{V_k(X_i)^4\} \leq 8c_0 \int \psi_{jk}^4(u)du \leq 2^{j+3}||\psi||_4^4 c_0, \tag{10.30}
$$

$$
E_{g_k}^n\{V_k(X_i)\} = 0. \tag{10.31}
$$

By the choice of j, there exists a constant $C > 0$ such that

$$2^j \geq C \left(\frac{n}{\log n} \right)^{\frac{1}{2\left(s+\frac{1}{2}-\frac{1}{r}\right)}},$$

and therefore, for n large enough,

$$j \log 2 \geq \frac{1}{2\left(s+\frac{1}{2}-\frac{1}{r}\right)} [\log n - \log \log n] + \log C \geq \lambda_n,$$

where $\lambda_n = \frac{\log n}{4\left(s+\frac{1}{2}-\frac{1}{r}\right)}$. Since card $R_j = 2^j$, we get also card $R_j \geq \exp(\lambda_n)$.
On the other hand, from (10.28) we deduce:

$$v_n^k \leq t_0^2 c_0 \log n \leq \lambda_n$$

for t_0 small enough. This yields conditions (ii) and (iv) of Lemma 10.1. To obtain the condition (iii) of Lemma 10.1, we must prove that $P_{g_k}^n(z_n^k > 0) \geq \pi_0 > 0$. This will follow from the next facts:

1.° $\zeta_n / \sqrt{\text{Var}\{\zeta_n\}}$ converges in $P_{g_k}^n$ distribution to a zero-mean normal variable with variance 1.

2.° $\text{Var}\{\zeta_n\} \geq \frac{c_0}{2} t_0^2 \log n \, (\geq 1$, say, for n large enough$)$.

3.° η_n converges to 0 in $P_{g_k}^n$ probability.

To prove 1° we apply the Central Limit Theorem with Lyapunov conditions (see for instance Pollard (1984)) and use (10.29). Next, to show 2°, note that, for n large enough, $\text{Var}\{\zeta_n\} = c_0^2 t_0^2 \log n \int \frac{\psi_{jk}^2(u)}{g_k(u)} du \geq (c_0/2) t_0^2 \log n \int \psi_{jk}^2(u) du$. The proof of 3° uses (10.29) and (10.30) and it is left to the reader.

Finally, applying Lemma 10.1 and the Markov inequality, we obtain:

$$R_n\left(\tilde{B}(s,r,q),p\right) \geq \left(\frac{\delta}{2}\right)^p \frac{\pi_0}{2}$$

with

$$\delta \simeq \sqrt{\frac{\log n}{n}} \left(\sqrt{\frac{n}{\log n}}\right)^{\frac{\frac{1}{2}-\frac{1}{p}}{s+\frac{1}{2}-\frac{1}{r}}} = \left(\frac{\log n}{n}\right)^{\alpha_2}.$$

This gives the result (10.20)-(10.21) in the sparse case.

Risk bounds: regular case

The regular case is characterized by the condition $r > p/(2s + 1)$. For the proof we use a more classical tool: Assouad's cube (Assouad (1983), Bretagnolle & Huber (1979)).

Let g_0, ψ_{jk} and R_j be as in the proof for the sparse case. As previously, denote by S_j the cardinality of R_j. Let $\varepsilon = (\varepsilon_1 \ldots \varepsilon_{S_j}) \in \{-1, +1\}^{S_j}$, and take $g^\varepsilon = g_0 + \gamma \sum_{k \in R_j} \varepsilon_k \psi_{jk}$. Let us denote by $\mathcal{G}$ the set of all such g^ε. Note that card $\mathcal{G}$ is of order 2^{2^j}. As $\int \psi_{jk} = 0$ (see (5.18)), we have $\int g^\varepsilon = 1$. Now, $\mathcal{G}$ is included in $\tilde{B}(s, r, q)$ if $\gamma \leq c_0 2^{-\frac{j}{2}} / \|\psi\|_\infty$ and $\|g^\varepsilon\|_{srq} \leq L$.

In view of Corollary 9.1,

$$\|g^\varepsilon\|_{srq} \leq \|g_0\|_{srq} + c_1 \gamma 2^{j(s+\frac{1}{2}-\frac{1}{r})} \Big(\sum_{k \in R_j} |\varepsilon_k|^r \Big)^{\frac{1}{r}}$$

where $c_1 > 0$ is a constant; we set for brevity $c_1 = 1$. Since $S_j = \text{card } R_j = 2^j$, we have $\|g^\varepsilon\|_{srq} \leq L$ if

$$2^{j(s+\frac{1}{2}-\frac{1}{r})} 2^{j/r} \gamma \leq \frac{L}{2}.$$

Thus, for large j only the following constraint on γ is necessary to guarantee that $g^\varepsilon \in \tilde{B}(s, r, q)$: $\gamma \leq (L/2) 2^{-j(s+\frac{1}{2})}$.

We now state a lemma which replaces Lemma 10.1 in this context.

LEMMA 10.2 *Let* $\delta = \inf_{\varepsilon \neq \varepsilon'} \|g^\varepsilon - g^{\varepsilon'}\|_p / 2$.

For ε *in* $\{-1, +1\}^{S_j}$, *put* $\varepsilon_{*k} = (\varepsilon_1' \ldots \varepsilon_{S_j}')$ *such that:*

$$\varepsilon_i' = \begin{cases} \varepsilon_i, & \text{if} \quad i \neq k, \\ -\varepsilon_i, & \text{if} \quad i = k. \end{cases}$$

If there exist $\lambda > 0$ *and* p_0 *such that*

$$P_{g\varepsilon}^n \big(\Lambda_n(g^{\varepsilon_{*k}}, g^\varepsilon) > e^{-\lambda} \big) \geq p_0, \forall \varepsilon, n,$$

then, for any estimator $\hat{f}$,

$$\max_{g_\varepsilon \in \mathcal{G}} E_{g^\varepsilon}^n \big[\|\hat{f} - g^\varepsilon\|_p^p \big] \geq \frac{S_j}{2} \delta^p e^{-\lambda} p_0.$$

Proof. Denote for the brevity $E_g = E_g^n$.

$$\max_{g_\varepsilon \in \mathcal{G}} E_{g^\varepsilon} \|\hat{f} - g^\varepsilon\|_p^p \geq \frac{1}{\text{card}\mathcal{G}} \sum_\varepsilon E_{g^\varepsilon} \|\hat{f} - g^\varepsilon\|_p^p$$

$$= \frac{1}{\text{card}\mathcal{G}} \sum_\varepsilon E_{g^\varepsilon} \int_a^b |\hat{f} - g^\varepsilon|^p(x) dx.$$

Let I_{jk} be the support of ψ_{jk}. As R_j is chosen so that those supports are disjoint, we have

$$\max_{g_\varepsilon \in \mathcal{G}} E_{g^\varepsilon} \|\hat{f} - g^\varepsilon\|_p^p \geq \frac{1}{\text{card}\mathcal{G}} \sum_\varepsilon E_{g^\varepsilon} \left[\sum_{k=1}^{S_j} \int_{I_{jk}} |\hat{f} - g^\varepsilon|^p(x) dx \right]$$

$$= \frac{1}{\text{card}\mathcal{G}} \sum_{k=1}^{S_j} \sum_\varepsilon E_{g^\varepsilon} \left[\int_{I_{jk}} |\hat{f} - g_0 + \varepsilon_k \gamma \psi_{jk}|^p(x) dx \right]$$

$$= \frac{1}{\text{card}\mathcal{G}} \sum_{k=1}^{S_j} \sum_{\substack{\varepsilon_i \in \{-1,+1\} \\ i \neq k}} \left\{ E_{g^\varepsilon} \left[\int_{I_{jk}} |\hat{f} - g_0 - \varepsilon_k \gamma \psi_{jk}|^p(x) dx \right] \right.$$

$$+ \left. E_{g^{\varepsilon \bullet k}} \left[\int_{I_{jk}} |\hat{f} - g_0 + \varepsilon_k \gamma \psi_{jk}|^p(x) dx \right] \right\}$$

$$= \frac{1}{\text{card}\mathcal{G}} \sum_{k=1}^{S_j} \sum_{\substack{\varepsilon_i \in \{-1,+1\} \\ i \neq k}} E_{g^\varepsilon} \left[\int_{I_{jk}} |\hat{f} - g_0 + \varepsilon_k \gamma \psi_{jk}|^p \right.$$

$$+ \left. \Lambda_n(g^{\varepsilon \bullet k}, g^\varepsilon) \int |\hat{f} - g_0 - \varepsilon_k \gamma \psi_{jk}|^p \right]$$

$$\geq \frac{1}{\text{card}\mathcal{G}} \sum_{k=1}^{S_j} \sum_{\substack{\varepsilon_i \in \{-1,+1\} \\ i \neq k}} E_{g^\varepsilon} \left[\delta^p I \left\{ \int_{I_{jk}} |\hat{f} - g_0 + \varepsilon_k \gamma \psi_{jk}|^p \geq \delta^p \right\} \right.$$

$$+ \left. \Lambda_n(g^{\varepsilon \bullet k}, g^\varepsilon) \delta^p I \left\{ \int_{I_{jk}} |\hat{f} - g_0 - \varepsilon_k \gamma \psi_{jk}|^p \geq \delta^p \right\} \right].$$

Remark that

$$\left(\int_{I_{jk}} |\hat{f} - g_0 + \varepsilon_k \gamma \psi_{jk}|^p \right)^{1/p} + \left(\int_{I_{jk}} |\hat{f} - g_0 - \varepsilon_k \gamma \psi_{jk}|^p \right)^{1/p} \geq \left(\int_{I_{jk}} |2\gamma \psi_{jk}|^p \right)^{1/p}$$

and

$$\left(\int_{I_{jk}} |2\gamma\psi_{jk}|^p\right)^{1/p} = ||g^\varepsilon - g^{\varepsilon \bullet k}||_p = \inf_{\varepsilon \neq \varepsilon'} ||g^\varepsilon - g^{\varepsilon'}||_p = 2\delta.$$

So we have

$$I\left\{\left(\int_{I_{jk}} |\hat{f} - g_0 + \varepsilon_k\gamma\psi_{jk}|^p\right)^{1/p} > \delta\right\} \geq I\left\{\left(\int_{I_{jk}} |\hat{f} - g_0 - \varepsilon_k\gamma\psi_{jk}|^p\right)^{1/p} \leq \delta\right\}.$$

We deduce that

$$\max_{g^\varepsilon \in \mathcal{G}} E_{g^\varepsilon}||\hat{f} - g^\varepsilon||_p^p \geq \frac{1}{\operatorname{card}\mathcal{G}}\sum_{k=1}^{S_j}\sum_{\substack{\varepsilon_i \in \{-1,+1\}\\ i \neq k}} \delta^p e^{-\lambda} P_{g^\varepsilon}\left(\Lambda_n(g^{\varepsilon\bullet k}, g^\varepsilon) \geq e^{-\lambda}\right)$$

$$\geq \frac{S_j}{2}\delta^p e^{-\lambda}p_0,$$

since $\operatorname{card}\mathcal{G} = 2^{S_j - 1}$.   □

It remains now to apply Lemma 10.2, i.e. to evaluate δ and $\Lambda_n(g^{\varepsilon\bullet k}, g^\varepsilon)$. Similarly to the calculations made for the sparse case, we write:

$$\Lambda_n(g^{\varepsilon\bullet k}, g^\varepsilon) = \prod_{i=1}^n\left(1 - \frac{2\frac{\gamma}{c_0}\varepsilon_k\psi_{jk}(X_i)}{1 + \frac{\gamma}{c_0}\varepsilon_k\psi_{jk}(X_i)}\right)$$

$$= \exp\left\{\sum_{i=1}^n\left[\frac{2\gamma}{c_0}V_k(X_i) - \frac{1}{2}\left(\frac{2\gamma}{c_0}V_k(X_i)\right)^2 + \kappa\left(\frac{2\gamma}{c_0}V_k(X_i)\right)\right]\right\}.$$

Define γ by $\frac{2\gamma}{c_0} = \frac{1}{\sqrt{n}}$. As in the sparse case proof, we show that

- $\frac{1}{\sqrt{n}}\sum_{i=1}^n V_k(X_i)/\sqrt{E_{g^\varepsilon}^n(V_k^2(X_i))}$ converges in $P_{g^\varepsilon}^n$ distribution to a variable $\mathcal{N}(0,1)$.

- $E_{g^\varepsilon}^n(V_k^2(X_i)) = c_0\int \frac{(\psi_{jk}(x))^2}{1 + \frac{\gamma}{c_0}\psi_{jk}(x)}dx \geq \frac{c_0}{2}$ since $\gamma\psi_{jk}(x) \leq c_0$ for n large enough.

- $\frac{1}{n}\sum_{i=1}^n\left[V_k^2(X_i) - E_{g^\varepsilon}^n(V_k^2(X_i))\right] \to 0$ as well as $\sum_{i=1}^N \kappa\left(\frac{1}{\sqrt{n}}V_k(X_i)\right) \to 0$, in $P_{g^\varepsilon}^n$ probability .

This entails the existence of $\lambda > 0$ and $p_0 > 0$ such that

$$P^n_{g\varepsilon} \left(\Lambda_n(g^{\varepsilon \bullet k}, g^\varepsilon) > e^{-\lambda} \right) \geq p_0.$$

It remains to evaluate δ. Since we need $\gamma \leq (L/2)2^{-j(s+1/2)}$ this leads to take $2^j \simeq n^{\frac{1}{1+2s}}$. Now

$$\delta = \inf_{\varepsilon \neq \varepsilon'} ||g^\varepsilon - g^{\varepsilon'}||_p/2 = ||\gamma \psi_{jk}||_p = \gamma 2^{j\left(\frac{1}{2} - \frac{1}{p}\right)} ||\psi||_p.$$

By substitution of this δ in the final inequality of Lemma 10.2 we obtain the result:

$$
\begin{aligned}
R_n\left(\tilde{B}(s,r,q), p\right) &\geq 2^{j-1} e^{-\lambda} p_0 \delta^p \\
&= 2^{-p-1} e^{-\lambda} p_0 ||\psi||_p^p c_0^p \left(\frac{1}{\sqrt{n}} 2^{j\left(\frac{1}{2} - \frac{1}{p}\right)} \right)^p 2^j \\
&\geq C n^{-\frac{sp}{2s+1}},
\end{aligned}
$$

where $C > 0$ is a constant.

From the sparse case computation we have

$$R_n\left(\tilde{B}(s,r,q), p\right) \geq C \left(\frac{\log n}{n} \right)^{\alpha_2 p} = C \left(\frac{\log n}{n} \right)^{\frac{\left(s - \frac{1}{r} + \frac{1}{p}\right)p}{2\left(s - \frac{1}{r}\right) + 1}}$$

where $C > 0$ is a constant. Thus

$$R_n\left(\tilde{B}(s,r,q), p\right) \geq C \max \left\{ \left(\frac{\log n}{n} \right)^{\alpha_2 p}, n^{-\frac{sp}{2s+1}} \right\},$$

which yields (10.20)–(10.21).

10.5   Asymptotic properties of wavelet thresholding estimates

The purpose of this section is to study the performance of L_p-risks of wavelet thresholding estimator f_n^* defined in (10.15) when the unknown density f belongs to a Besov class $\tilde{B}(s,r,q)$. Then we compare the result with the lower bound (10.20) of Theorem 10.3, and thus obtain an answer to Questions 10.1 and 10.5.

Let, as in Theorem 10.3,

$$\alpha_1 = \frac{s}{2s+1}, \quad \alpha_2 = \frac{s - \frac{1}{r} + \frac{1}{p}}{2(s - \frac{1}{r}) + 1},$$

and define

$$\alpha = \begin{cases} \alpha_1 \quad, & \text{if} \quad r > \frac{p}{2s+1}, \\ \alpha_2 \quad, & \text{if} \quad r \leq \frac{p}{2s+1}. \end{cases}$$

Suppose that the parameters j_0, j_1, t of the wavelet thresholding estimator (10.16) satisfy the assumptions:

$$2^{j_0(n)} \simeq \begin{cases} n^{\frac{\alpha}{s}} \quad, & \text{if} \quad r > \frac{p}{2s+1}, \\ n^{\frac{\alpha}{s}} (\log n)^{\alpha(p-r)/sr} \quad, & \text{if} \quad r \leq \frac{p}{2s+1}, \end{cases} \tag{10.32}$$

$$2^{j_1(n)} \simeq (n/\log n)^{\alpha/s'}, \tag{10.33}$$

$$t = t_j = c\sqrt{\frac{j}{n}}, \tag{10.34}$$

where $c > 0$ is a positive constant.

Note that the threshold t in (10.34) depends on j.

THEOREM 10.4 *Let $1 \leq r, q \leq \infty, 1 \leq p < \infty, s > 1/r$ and $r < p$, and let f_n^* be the estimator (10.15) such that:*

- *the father wavelet φ satisfies the conditions of Theorem 9.4 for some integer $N \geq 0$,*

- *$\beta_{jk}^* = \hat{\beta}_{jk}^H$ with the variable threshold $t = t_j = c\sqrt{\frac{i}{n}}$,*

- *the assumptions (10.32)-(10.34) are satisfied, and $s < N + 1$.*

Then, for $c > 0$ large enough, one has

$$\sup_{f \in \tilde{B}(s,r,q)} E\|f_n^* - f\|_p^p \leq \begin{cases} C(\log n)^\delta n^{-\alpha_1 p} & , \text{if } r > \frac{p}{2s+1}, \\ C(\log n)^{\delta'} \left(\frac{\log n}{n}\right)^{\alpha_2 p} & , \text{if } r = \frac{p}{2s+1}, \\ C\left(\frac{\log n}{n}\right)^{\alpha_2 p} & , \text{if } r < \frac{p}{2s+1}, \end{cases}$$

where δ and δ' are positive constants depending only on p, s, r, q, and $C > 0$ is a constant depending only on p, s, r, q, L, L'.

REMARK 10.5 • In the sparse case $r < \frac{p}{2s+1}$, the rate is sharp: Theorems 10.3 and 10.4 agree. The wavelet thresholding estimator attains the optimal rate of convergence $\left(\frac{\log n}{n}\right)^{\alpha_2}$.

• On the boundary $r = \frac{p}{2s+1}$ of the sparse zone the lower bound of Theorem 10.3 and the upper bound of Theorem 10.4 differ in a logarithmic factor. As this result can be compared with the result obtained in the Gaussian white noise setting, (Donoho, Johnstone, Kerkyacharian & Picard (1997)) the upper bound of Theorem 10.4 is likely to be correct whereas the lower bound (10.20) is too optimistic. In this boundary case the optimal rate for the Gaussian white noise setting turns out to depend on the parameter q (see Donoho et al. (1997)).

• In the regular case $r > \frac{p}{2s+1}$, the bounds of Theorem 10.3 and 10.4 still do not agree. In this case the logarithmic factor is an extra penalty for the chosen wavelet thresholding. However, it can be proved, that the logarithmic factor can be removed by selecting a slightly different threshold: $t_j = c\sqrt{\frac{j-j_0}{n}}$ (Delyon & Juditsky (1996a)).

REMARK 10.6 It has been proved in Corollary 10.3 that if $r \geq p$, then

$$R_n^{lin}\left(\tilde{B}(s,r,q),p\right) \simeq R_n\left(\tilde{B}(s,r,q),p\right).$$

From 10.22 and Theorem 10.4 we see that, for $r < p$, we have strict inequalities:

$$R_n^{lin}\left(\tilde{B}(s,r,q),p\right) >> R_n\left(\tilde{B}(s,r,q),p\right).$$

REMARK 10.7 The constant $c > 0$ in the definition of the threshold (10.34) can be expressed in terms of s, r, q, L, and it does not depend on j, n and on a particular density f. We do not discuss here why the particular form (10.34) of $t = t_j$ is chosen: the discussion is deferred to Chapter 11.

REMARK 10.8 The assumption on φ in Theorem 10.4 is rather general. For example, it is satisfied if φ is bounded, compactly supported and the derivative $\varphi^{(N+1)}$ is bounded. These conditions hold for the usual bases of compactly supported wavelets (Daubechies, coiflets, symmlets) of a sufficiently high order (see Chapter 7).

Summarizing the results of Theorems 10.3 - 10.4, and the Remarks 10.5–10.6, we are now able to answer the Questions 10.1 and 10.5:

- *Optimal rates of convergence on the Besov classes $\tilde{B}(s,r,q)$ are*

  - $n^{-\frac{s}{2s+1}}$ in the regular case $(r > \frac{p}{2s+1})$,

  - $\left(\frac{\log n}{n}\right)^{\frac{s-\frac{1}{r}+\frac{1}{p}}{2(s-\frac{1}{r})+1}}$ in the sparse case $(r < \frac{p}{2s+1})$.

  - There is an uncertainty on the boundary $r = \frac{p}{2s+1}$, where the optimal rate is $n^{-s/(2s+1)}$, to within some logarithmic factor (the problem of defining this factor remains open).

- *The properly thresholded wavelet estimator (10.15) attains the optimal rates (in some cases to within a logarithmic factor).*

The proof of Theorem 10.4 can be found in Donoho, Johnstone, Kerkyacharian & Picard (1996). We do not reproduce it here, but rather consider a special case where the bound on the risk of a wavelet thresholding estimator f_n^* is simpler. This will allow us to present, without excessive technicalities, the essential steps of the proof.

Assume the following

$$p = 2, \ 1 \leq r = q < 2, \ s > \frac{1}{r}, \tag{10.35}$$

$$
\begin{aligned}
2^{j_0} &\simeq n^{\frac{1}{2s+1}}, \\
2^{j_1} &\geq n^{\alpha_1/(s-\frac{1}{r}+\frac{1}{2})},
\end{aligned}
\tag{10.36}
$$

$$t = c\sqrt{\frac{\log n}{n}}, \tag{10.37}$$

for some large enough $c > 0$.

Under the condition (10.35), clearly, $p > r > \frac{p}{2s+1}$. Thus, we are in the intermediate zone (see Figure 10.13), and the lower bound on the minimax risk is, in view of Theorem 10.3,

$$r_n(s,r,p,q) = r_n(s,r,2,r) = n^{-\frac{2s}{2s+1}}.$$

The next proposition shows, that, to within a logarithmic factor, the asymptotic behavior of the wavelet thresholding estimator (10.15) is of the same order.

PROPOSITION 10.3 *Let f_n^* be the estimator (10.15) such that:*

- *the father wavelet φ and the mother wavelet ψ are bounded and compactly supported, and for some integer $N \geq 0$, the derivative $\varphi^{(N+1)}$ is bounded,*

- *$\beta_{jk}^* = \hat{\beta}_{jk}^H$, with the threshold $t = c\sqrt{\frac{\log n}{n}}$,*

- *the assumptions (10.35) – (10.37) are satisfied, and $s < N + 1$.*

Then, for $c > 0$ large enough, one has

$$\sup_{f \in \tilde{B}(s,r,r)} E\|f_n^* - f\|_2^2 \leq C(\log n)^\gamma R_n(\tilde{B}(s,r,r),2) \simeq (\log n)^\gamma n^{-\frac{2s}{2s+1}},$$

where $\gamma = 1 - \frac{r}{2}$, and $C > 0$.

Proof Observe first that the choice of the threshold $t = c\sqrt{\frac{\log n}{n}}$ instead of $t_j = c\sqrt{\frac{j}{n}}$ does not make a big difference since for $j_0 \leq j \leq j_1$ there exist two constants c_1 and c_2 such that $c_1\sqrt{\frac{\log n}{n}} \leq \sqrt{\frac{j}{n}} \leq c_2\sqrt{\frac{\log n}{n}}$. This will be used at the end of the proof.

Observe also that the functions $f \in \tilde{B}(s,r,r)$ are uniformly bounded:

$$\|f\|_\infty \leq C_*,$$

where $C_* > 0$ depends only on s, r, L. This is a consequence of the (compact) embedding of $B_r^{sr}(I\!R)$ into $C(I\!R)$ for $s > 1/r$ (Corollary 9.2 (iv)). As before, we use the generic notation C for positive constants, possibly different. We shall also write f^* for f_n^*. Note that $\hat{f}_{j_0-1}(x) = \sum_k \alpha_{j_0 k} \varphi_{j_0 k}(x)$ (cf.(10.1)). By orthogonality of the wavelet basis, one gets

$$E\|f^* - f\|_2^2 = E\|\hat{f}_{j_0-1} - E(\hat{f}_{j_0-1})\|_2^2$$

$$+ \sum_{j=j_0}^{j_1} \sum_{k \in \Omega_j} \left(E[(\hat{\beta}_{jk} - \beta_{jk})^2 I\{|\hat{\beta}_{jk}| > t\}] + \beta_{jk}^2 P\{|\hat{\beta}_{jk}| \leq t\} \right)$$

$$+ \sum_{j=j_1}^{\infty} \sum_{k \in \Omega_j} \beta_{jk}^2 = T_1 + T_2 + T_3 + T_4, \tag{10.38}$$

where $\Omega_j = \{k : \beta_{jk} \neq 0\}$. Let us observe that card $\Omega_j \leq 2^j L' + \tau$, where τ is the maximum of the lengths of the supports of φ and ψ (cf. Remark 10.1).

The terms T_j are estimated as follows. First, using Proposition 10.1 and (10.36), we get

$$
\begin{aligned}
T_1 &= E\|\hat{f}_{j_0-1} - E(\hat{f}_{j_0-1})\|_2^2 \leq C\frac{2^{j_0}}{n} \\
&\leq Cn^{-\frac{2s}{2s+1}}.
\end{aligned}
\tag{10.39}
$$

Using the parts (i) and (iii) of Corollary 9.2, we obtain $B_r^{sr}(\mathbb{R}) \subset B_2^{s'2}(\mathbb{R})$, for $r < 2$, where $s' = s - \frac{1}{r} + \frac{1}{2}$. Thus, any function f that belongs to the ball $\tilde{B}(s, r, r)$ in $B_r^{sr}(\mathbb{R})$, also belongs to a ball in $B_2^{s'2}(\mathbb{R})$. Therefore, by Theorem 9.6, the wavelet coefficients β_{jk} of f satisfy the condition (B3):

$$
\sum_{j=0}^{\infty} 2^{2js'} \sum_k \beta_{jk}^2 < \infty.
$$

Hence,

$$
\begin{aligned}
T_4 &= \sum_{j=j_1}^{\infty} \sum_{k \in \Omega_j} \beta_{jk}^2 \leq C2^{-2j_1 s'} \sum_{j=0}^{\infty} 2^{2js'} \sum_k \beta_{jk}^2 \\
&\leq Cn^{-\frac{2s}{2s+1}},
\end{aligned}
\tag{10.40}
$$

where we again use (10.36).

To estimate the terms T_2 and T_3 write

$$
\begin{aligned}
T_2 &= \sum_{j=j_0}^{j_1} \sum_{k \in \Omega_j} E((\hat{\beta}_{jk} - \beta_{jk})^2)[I\{|\hat{\beta}_{jk}| > t, |\beta_{jk}| > \frac{t}{2}\} \\
&\quad + I\{|\hat{\beta}_{jk}| > t, |\beta_{jk}| \leq \frac{t}{2}\}], \\
T_3 &= \sum_{j=j_0}^{j_1} \sum_{k \in \Omega_j} \beta_{jk}^2[P\{|\hat{\beta}_{jk}| \leq t, |\beta_{jk}| \leq 2t\} \\
&\quad + P\{|\hat{\beta}_{jk}| \leq t, |\beta_{jk}| > 2t\}].
\end{aligned}
$$

Note that

$$
I\{|\hat{\beta}_{jk}| > t, |\beta_{jk}| \leq \frac{t}{2}\} \leq I\{|\hat{\beta}_{jk} - \beta_{jk}| > \frac{t}{2}\},
\tag{10.41}
$$

$$
I\{|\hat{\beta}_{jk}| \leq t, |\beta_{jk}| > 2t\} \leq I\{|\hat{\beta}_{jk} - \beta_{jk}| > \frac{t}{2}\},
$$

and, if $|\hat{\beta}_{jk}| \le t$, $|\beta_{jk}| > 2t$, then $|\hat{\beta}_{jk}| \le \frac{|\beta_{jk}|}{2}$, and $|\hat{\beta}_{jk} - \beta_{jk}| \ge |\beta_{jk}| - |\hat{\beta}_{jk}| \ge |\beta_{jk}|/2$. Therefore

$$\beta_{jk}^2 \le 4(\hat{\beta}_{jk} - \beta_{jk})^2. \tag{10.42}$$

Using (10.41) and (10.42), we get

$$
\begin{aligned}
T_2 + T_3 \quad \le \quad & \sum_{j=j_0}^{j_1} \sum_{k \in \Omega_j} \left[E\left\{ (\hat{\beta}_{jk} - \beta_{jk})^2 \right\} I\{|\beta_{jk}| > \frac{t}{2}\} \right. \\
& + \left. \beta_{jk}^2 I\{|\beta_{jk}| \le 2t\} \right] \\
& + 5 \sum_{j=j_0}^{j_1} \sum_{k \in \Omega_j} E\left\{ (\hat{\beta}_{jk} - \beta_{jk})^2 I\{|\hat{\beta}_{jk} - \beta_{jk}| > \frac{t}{2}\} \right\}. \tag{10.43}
\end{aligned}
$$

Clearly,

$$E(\hat{\beta}_{jk} - \beta_{jk})^2 = \frac{1}{n} Var\{\psi_{jk}(X_1)\} \le \frac{1}{n} \int \psi_{jk}^2(x) f(x) dx \le \frac{1}{n} \|f\|_\infty \le \frac{C_*}{n}.$$

Also, using the Markov inequality, one easily gets,

$$\text{card } \{(j,k) : j_0 \le j \le j_1, |\beta_{jk}| > \frac{t}{2}\} \le \left(\frac{2}{t}\right)^r \sum_{j=j_0}^{j_1} \sum_k |\beta_{jk}|^r.$$

This yields:

$$
\begin{aligned}
\sum_{j=j_0}^{j_1} \sum_{k \in \Omega_j} \quad & E\left((\hat{\beta}_{jk} - \beta_{jk})^2 \right) I\{|\beta_{jk}| > \frac{t}{2}\} \\
& \le \frac{C}{n} \left(\frac{2}{t}\right)^r 2^{-j_0 r(s+\frac{1}{2}-\frac{1}{r})} \sum_{j=j_0}^{j_1} \sum_k 2^{jr(s+\frac{1}{2}-\frac{1}{r})} |\beta_{jk}|^r \\
& \le \frac{C}{n} \left(\frac{n}{\log n}\right)^{r/2} 2^{-j_0 r(s+\frac{1}{2}-\frac{1}{r})} \\
& \le C n^{-\frac{2s}{2s+1}}, \tag{10.44}
\end{aligned}
$$

where we used (10.35), (10.36) and the condition

$$\sum_{j=0}^{\infty} \sum_k 2^{jr(s+\frac{1}{2}-\frac{1}{r})} |\beta_{jk}|^r \le C \tag{10.45}$$

that follows from the fact that $f \in \tilde{B}(s,r,r)$ and from Theorem 9.6.

Next, as $r < 2$,

$$\sum_{j=j_0}^{j_1} \sum_k \beta_{jk}^2 I\{|\beta_{jk}| \le 2t\}$$

$$\le (2t)^{2-r} \sum_{j=j_0}^{j_1} \sum_k |\beta_{jk}|^r \le C \left(\frac{\log n}{n} \right)^{\frac{2-r}{2}} 2^{-j_0 r(s+\frac{1}{2}-\frac{1}{r})}$$

$$\le C n^{-\frac{2s}{2s+1}} (\log n)^{\frac{2-r}{2}}, \tag{10.46}$$

where (10.45) was used.

Define T_{41} as the last term in (10.43). Elementary calculation shows:

$$
\begin{aligned}
E\left\{ (\hat{\beta}_{jk} - \beta_{jk})^4 \right\} &\le \frac{C}{n^2} E\left\{ \psi_{jk}^4(X_1) \right\} \\
&= \frac{C}{n^2} \int \psi_{jk}^4(x) f(x) dx \\
&\le \frac{C}{n^2} \|f\|_\infty \int \psi_{jk}^4(x) dx \le \frac{C 2^j}{n^2}.
\end{aligned}
$$

Using this and the Cauchy-Schwarz inequality, one obtains

$$
\begin{aligned}
T_{41} &= 5 \sum_{j=j_0}^{j_1} \sum_{k \in \Omega_j} E\left\{ (\hat{\beta}_{jk} - \beta_{jk})^2 I\{|\hat{\beta}_{jk} - \beta_{jk}| > \frac{t}{2}\} \right\} \\
&\le \left(\frac{C}{n^2} \right)^{1/2} \sum_{j=j_0}^{j_1} \sum_{k \in \Omega_j} 2^{j/2} P^{1/2}\{|\hat{\beta}_{jk} - \beta_{jk}| > \frac{t}{2}\} \\
&\le \frac{C}{n} \sum_{j=j_0}^{j_1} \sum_{k \in \Omega_j} 2^{j/2} P^{1/2}\{|\hat{\beta}_{jk} - \beta_{jk}| > c\sqrt{\frac{j}{n}}\}, \tag{10.47}
\end{aligned}
$$

where (10.36) and (10.37) were used.

The last probability in (10.47) is evaluated using the following well known lemma (see the proof in Appendix C).

LEMMA 10.3 (Bernstein's inequality.) *Let $\zeta_1, \ldots, \zeta_n$ be i.i.d. bounded random variables, such that $E(\zeta_i) = 0$, $E(\zeta_i^2) \le \sigma^2$, $|\zeta_i| \le \|\zeta\|_\infty < \infty$. Then*

$$P(|\frac{1}{n}\sum_{i=1}^n \zeta_i| > \lambda) \le 2 \exp\left(-\frac{n\lambda^2}{2(\sigma^2 + \|\zeta\|_\infty \lambda/3)} \right), \quad \forall \lambda > 0.$$

Applying Lemma 10.3 to $\zeta_i = \psi_{jk}(X_i) - E(\psi_{jk}(X_i))$, and noting that one can define $\sigma^2 = C_* \geq ||f||_\infty \geq Var\{\psi_{jk}(X_1)\}$, we conclude that, if $c > 0$ is large enough,

$$P\left\{|\hat{\beta}_{jk} - \beta_{jk}| > c\sqrt{\frac{j}{n}}\right\} \leq 2^{-4j}.$$

Next, substitute this into (10.47), and obtain the following

$$T_{41} \leq \frac{C}{n}\sum_{j=j_0}^{j_1}\sum_{k\in\Omega_j} 2^{-3j/2} \leq \frac{C}{n}\sum_{j=j_0}^{j_1} 2^{-j/2}$$

$$\leq \frac{C}{n}2^{-j_0/2} \leq \frac{C}{n}, \tag{10.48}$$

where we used the fact that $\mathrm{card}\,\Omega_j \leq C2^j$, mentioned at the beginning of the proof.

To end the proof of the proposition it remains to put together (10.38) – (10.40), (10.43), (10.44) and (10.46) – (10.48). □

10.6 Some real data examples

Estimation of financial return densities

For a given time series of financial data S_i (e.g. stock prices), returns are defined as the first differences of the log series, $X_i = \log S_i - \log S_{i-1}$. A basic distributional assumption in the statistical analysis of finance data is that returns are approximately normally distributed. The assumption is helpful in applying the maximum likelihood rule for certain models e.g. the ARCH specification (Gourieroux 1992). Another reason for the dominance of the normality assumption in finance is that in traditional equilibrium models as the capital asset pricing model (CAPM), established by Sharpe (1964) and Lintner (1965), utility functions are quadratic. Thus they only depend on the first two moments of the return distribution. Also in option pricing the normality assumption of returns together with constant volatility (variance) of X_i is vital. The Black & Scholes (1973) formula yields under this assumption a unique option price as a function of strike price and volatility.

It has been criticized in the recent literature that the normality assumption does not capture typical phenomena of the distribution of financial data like foreign exchange or stock returns: thickness of tails, slim center concentration, multimodality or skewness for different market periods, Gourieroux (1992).

Here we apply wavelet density estimators to analyze the normality versus non-normality issue in two examples. Note that we put ourselves here into the framework of dependent data X_i. Results similar to thos formulated above hold for this framework as well (see Tribouley & Viennet (1998)). For the first example, we consider the data given in Fama (1976, Table 4.1, p.102). It contains the returns of IBM stocks from July 1963 - June 1968 and the returns of an equally weighted market portfolio. Our interest is in comparing the distributions of these two data sets.

Figure 10.14 contains the IBM data, a parametric normal density estimate, the wavelet estimator with soft thresholding of $0.6 \max |\hat{\beta}_{jk}|$, $j_1 = 4$, for symmlet $S4$ and a kernel estimate. The soft threshold was determined by visual inspection.

The normal density estimator was computed with the mean and standard deviation of the return data plugged into a normal density. The kernel density estimate with a quartic kernel is marked as a dashed curve. The nonnormality is clearly visible in the wavelet estimate and corresponds to different market periods, Fama (1976). The normal density estimator cannot capture the local curvature of this data.

Consider next the second data set of Fama (1976), related to the equally weighted market portfolio. We choose the same threshold level as for the IBM data. It can be seen from Figure 10.15 (threshold value $0.6 \max_{j,k} |\hat{\beta}_{jk}|$) that the estimate is closer to a normal density than for the IBM data. This fits well with the intuitive hypothesis that the portfolio (which is the average of many stock elements) would have a quasi-Gaussian behavior.

We turn now to the second example related to the data set of Section 11. The series of exchange rate values DEMUSD (DM to US dollar) is given in the upper half of Figure 10.16. The time period of observations here is the same as in bid-ask speeds of Figure 1.1 (Section 1.1). The corresponding returns density is displayed in the lower half. The feature of thick tails together with a very concentrated slim center peak is clearly visible. The normal distribution density underestimates the central peak and has higher tails outside the one standard deviation region. Based on this observation recent literature in the analysis of this data proposes Pareto distribution

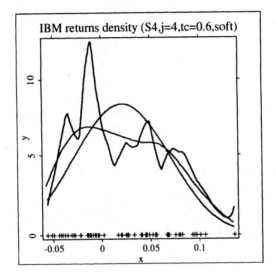

Figure 10.14: Density estimate of IBM returns. Soft thresholding, $t = 0.6 \max_{j,k} |\hat{\beta}_{jk}|$.

densities for example.

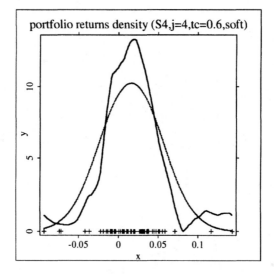

Figure 10.15: Density estimate of equally weighted portfolio. Soft thresholding, $t = 0.6 \max_{j,k} |\hat{\beta}_{jk}|$.

Estimation of income densities

The Family Expenditure Survey (FES) is based on a representative sample of private households in the United Kingdom in every year since 1957. The sample size of the FES is approximately 7000 households per year, which amount to about 5 percent of all households in the United Kingdom. The FES contains detailed information on household characteristics, like household size and composition, occupation, age, etc. The theory of market demand as described by Hildenbrand (1994) concentrates on the analysis of the structure of income.

A feature important for the application of the economic theory is the stability of income distribution over time. We consider this question by estimating the densities of the FES for the years 1969 - 1983. Earlier approaches have been based on a log-normality assumption of the income distribution,

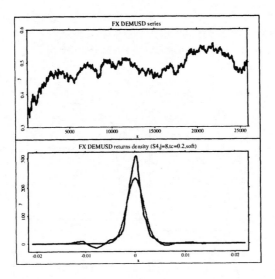

Figure 10.16: A comparison of density estimates. DEMUSD spot rates in upper graph; normal and wavelet estimates in lower graph.

described in Hildenbrand (1994). This parametric assumption though does not allow for the possible changes in income that have been observed especially during the Thatcher era. In particular, the possibility of multimodality is explicitly excluded.

The densities were estimated with a symmlet $S4$ wavelet and soft thresholding of $t = 0.1 \max_{j,k} |\hat{\beta}_{jk}|$, based on 256 bins computed from the about 7000 observations per year. Figure 10.17 shows the density estimates for the first four years 1969 - 1972. These and the following density estimates have been computed from normalized income, i.e. the observations were divided by their mean. The mean of income each year is thus normalized to be equal to 1. The first two years are unimodal and left skew densities whereas the density for 1971 show a pronounced shoulder in the region of 80 percent mean income. This effect vanishes for the 1972 but reappears in Figure 10.18 for 1973 and 1975. The higher peak near the mean income which is a continuous structural feature for the first 8 years diminishes over the next 7 years. Figure 10.19 shows two unimodal densities and then a shift in magnitude of the two modes which is continued until 1983, see Figure 10.20. The collection of

all 15 densities is displayed in the lower right of Figure 10.20. We conclude
from our nonparametric wavelet analysis for these curves that there has been
a shift in the income distribution from the peak at about $x = 1$ to the lower
level $x = 0.8$.

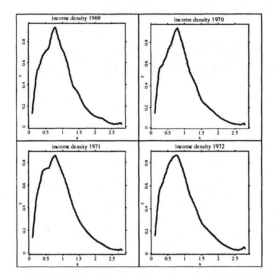

Figure 10.17: FES Income densities 1969-1972.

10.7   Comparison with kernel estimates

Kernel density estimates have a long tradition in data smoothing. It is therefore interesting to compare the wavelet estimates with kernel estimates. A
kernel density estimator $\hat{f}_h$ is defined via a kernel K and a bandwidth h, see
e.g. Silverman (1986),

$$\hat{f}_h(x) = n^{-1}h^{-1} \sum_{i=1}^{n} K\left(\frac{x - X_i}{h}\right). \tag{10.49}$$

In application of (10.49) we need to select a bandwidth and a kernel K. We
applied the two methods to $n = 500$ data points with density

$$f(x) = 0.5\varphi(x) + 3\varphi\{10(x - 0.8)\} + 2\varphi\{10(x - 1.2)\} \tag{10.50}$$

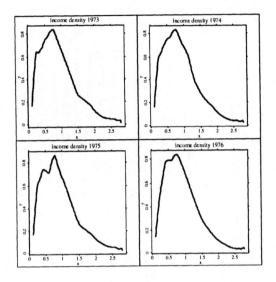

Figure 10.18: FES Income densities 1973-1976.

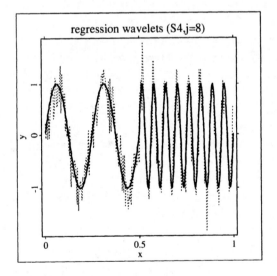

Figure 10.19: FES Income densities 1977-1980.

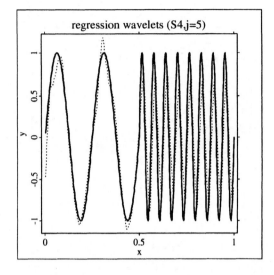

Figure 10.20: FES Income densities 1981-1983, 1969-1983.

Here φ denotes the standard normal density. A diagram of the density to-gether with the data is shown in Figure 10.21.

We have investigated seven different bandwidth choice methods as in Park & Turlach (1992). Table 10.3 below gives the values h suggested by these methods for the Gaussian kernel $K = \varphi$ and the Quartic kernel $K(u) = \frac{15}{16}(1 - u^2)^2 \, I\{|u| \leq 1\}$.

In Figure 10.22 we show two different kernel density estimators with bandwidths $h = 0.18$ and $h = 0.6$ (dotted line), respectively. The computation was done with the Quartic kernel. One sees the basic problem of the kernel estimate: the bandwidth is either too small or too high. The left shoulder is well estimated by the kernel estimate with bandwidth $h = 0.6$ but the two peaks are not picked up. The smaller bandwidth estimate models the peaks nicely but fails on the shoulder part.

In comparison with the hard thresholded wavelet density estimator of Figure 10.23 the kernel estimates are unfavorable. The wavelet density estimator was computed with the highest level $j_1 = 8$ (dotted line). The threshold was set to 0.4 of the maximal value. The kernel density estimate was taken with "medium" bandwidth $h = 0.4$, see Table 10.3. The wavelet density estimate

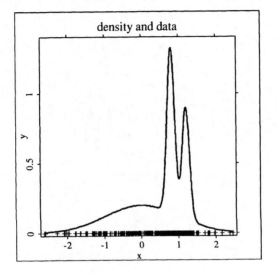

Figure 10.21: A trimodal density and $n = 500$ data points.

Method	K=Gauss	K=Quartic
Least squares cross validation	0.067	0.175
Biased cross validation	0.4	1.049
Smoothed cross validation	0.387	1.015
Bandwidth factorized cross validation	0.299	0.786
Park and Marron plug in	0.232	0.608
Sheather and Jones plug in	0.191	0.503
Silverman's rule of thumb	0.45	1.18

Table 10.3: Different bandwidth selectors for data of Figure 10.21

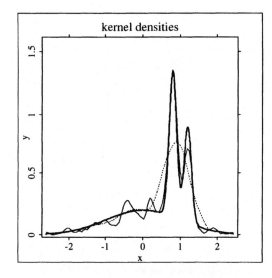

Figure 10.22: The density with two kernel density estimates.

captures the right peak partly and is more stable on the left shoulder side. This performance is even improved for the soft thresholded wavelet density estimator, see Figure 10.24. The peaks are both well represented and except for a small trough the wavelet density estimate is remarkably stable in the interval $[-3, 0]$.

The integrated squared error (ISE) for the kernel estimate $\hat{f}_h$ was 0.019 whereas the wavelet estimate resulted in a value of $ISE = 0.0099$ (hard) and of 0.0063 (soft).

In summary we can say that this small study of comparison has shown what was expected. Kernel density estimators are not locally adaptive, unless we employ a more complicated local bandwidth choice. Wavelet estimators are superior but may show some local variability as in Figure 10.24 for example. For data analytic purposes with small to moderate data size a kernel estimate may be preferred for its simplicity and wide distribution. For finer local analysis and good asymptotic properties the wavelet estimator is certainly the method to be chosen.

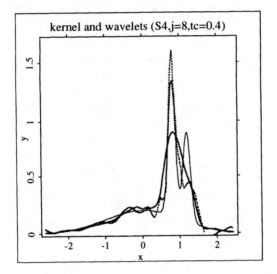

Figure 10.23: The density, a kernel estimate and a wavelet estimate with hard thresholding ($S4$, $j_1 = 8$, $t = 0.4 \max_{j,k} |\hat{\beta}_{jk}|$).

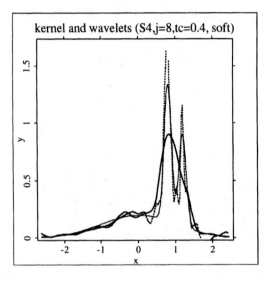

Figure 10.24: The density, a kernel estimate and a wavelet estimate with soft thresholding ($S4$, $j_1 = 8$, $t = 0.4 \max_{j,k} |\hat{\beta}_{jk}|$).

10.8 Regression estimation

Assume that
$$Y_i = f(X_i) + \xi_i, \ i = 1, \dots, n,$$

where ξ_i are independent random variables, $E(\xi_i) = 0$, and X_i are on the regular grid in the interval $[0,1]$: $X_i = \frac{i}{n}$. Consider the problem of estimating f given the data $(Y_1, \dots, Y_n)$.

The *linear wavelet regression estimator* $\hat{f}_{j_1}$ for f is defined by (10.1), with a different definition of the estimated coefficients α_{jk}, β_{jk}:

$$\hat{\alpha}_{jk} = \frac{1}{n} \sum_{i=1}^{n} Y_i \, \varphi_{jk}(X_i), \tag{10.51}$$

$$\hat{\beta}_{jk} = \frac{1}{n} \sum_{i=1}^{n} Y_i \, \psi_{jk}(X_i). \tag{10.52}$$

This choice of $\hat{\alpha}_{jk}$ and $\hat{\beta}_{jk}$ is motivated by the fact that (10.51) and (10.52) are "almost" unbiased estimators of α_{jk} and β_{jk} for large n. For example,

$$E(\hat{\beta}_{jk}) = \frac{1}{n} \sum_{i=1}^{n} f(\frac{i}{n}) \psi_{jk}(\frac{i}{n}) \approx \int f \psi_{jk}$$

if f and ψ are smooth enough and ψ satisfies the usual assumptions, see Remark 10.1.

The *wavelet thresholding regression estimator* f_n^* is defined by (10.15) and (10.13), (10.14), respectively, for soft and hard thresholding, with $\hat{\alpha}_{jk}$ and $\hat{\beta}_{jk}$ as in (10.51), (10.52).

The remarks concerning the choice of parameters j_0, j_1, the functions φ and ψ and thresholding (see Sections 10.2 – 10.4) remain valid here.

It is important that the points X_i are on the regular grid in the interval $[0,1]$. One should change the definition of the estimators otherwise. This is discussed for example by Hall & Turlach (1995), Hall, McKay & Turlach (1996), Neumann & Spokoiny (1995), and we would like to dwell a little more on it here.

Different techniques can be implemented. The first technique is based on a preliminary binning and scaling of the observation interval to map it into [0,1], and it is close to WARPing, see Härdle & Scott (1992). We implement this technique in the simulations below. The idea of the construction is simular to that of (10.10) - (10.12). We first compute a regressogram estimator with bins of width Δ centered at equispaced gridpoint $z_1, \ldots, z_m$. For computational reasons (to make possible the use of discrete wavelet transform, see Chapter 12), it is necessary to choose m as a power of 2: $m = 2^K$, where $K \geq j_1$ is an integer. Here Δ should be a very small number (in relative scale). Let $\hat{y}_1, \ldots, \hat{y}_m$ be the values of the regressogram at gridpoints $z_1, \ldots, z_m$:

$$\hat{y}_i = \frac{\sum_{s=1}^n Y_s I\{|X_s - z_i| \leq \Delta/2\}}{\sum_{s=1}^n I\{|X_s - z_i| \leq \Delta/2\}}, \quad i = 1, \ldots, m.$$

Next, we apply the formulas (10.10) - (10.12) to get the values f_l of the regression estimator at gridpoints $z_1, \ldots, z_m$.

The second technique of handling the non-equispaced case was proposed by Neumann & Spokoiny (1995). It is related to the Gasser-Müller kernel regression estimator, see Härdle (1990, Section 3.2). The computation of this estimator seems to be more difficult than that of the binned one since it cannot in general be reduced to the discrete wavelet transform algorithm.

Note that, as we work on the bounded interval and not on $I\!\!R$, the wavelet base $\{\varphi_{j_0 k}, \psi_{jk}\}$ is no longer an ONB. In practice this will appear as boundary effects near the endpoints of the interval [0, 1]. Several ways of correction are possible. First, the implementation of wavelet orthonormal bases on the interval as in Meyer (1991) and Cohen, Daubechies & Vial (1993). A second approach would be a standard boundary correction procedure as in Härdle (1990), based on boundary kernels. A third approach presented later in this section is based on mirroring.

Let us first consider wavelet regression smoothing without boundary correction. The wavelet technique for regression is applied to the data in Figure 10.25. We generated the function

$$f(x) = \sin(8\pi x)I\{x \leq 1/2\} + \sin(32\pi x)I\{x > 1/2\}, \quad x \in (0, 1) \qquad (10.53)$$

with normal noise ξ_i whose standard deviation is 0.4. The 512 observations are shown as plus signs, and the true function is displayed as a solid line.

This example is the same as in Figures 1.12, 1.13 but we have added observation noise. Figure 10.26 shows the linear wavelet estimator $\hat{f}_{j_1}$ with

$S4$ father and mother wavelets, $j_0 = 0$ and $j_1 = 8$: the estimator goes almost through the observation points.

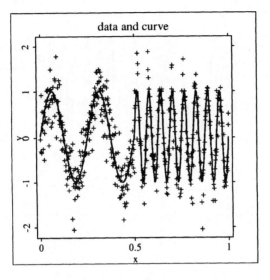

Figure 10.25: Data and regression curve.

Next we restrict the levels to a maximum of $j_1 = 5$ and start with $j_0 = 0$. The resulting linear estimate is given in Figure 10.27. The power of wavelet smoothing again becomes apparent: the high frequencies are well modelled and at the same time the lower frequencies in the left half of the observation interval are nicely represented.

Wavelet thresholding regression estimators are defined by (10.13)–(10.15), with the empirical wavelet coefficients given in (10.51), (10.52). We briefly discuss their performance on the same example as considered above in this section.

Hard thresholding with $t = 0.2 \max |\hat{\beta}_{jk}|$ gave about the same ISE as soft thresholding. We therefore show only the soft thresholding estimate in Figure 10.28.

Observe that the estimator behaves quite reasonably at the endpoints of the interval. Boundary correction in this example, at least visually, turns out not to be necessary.

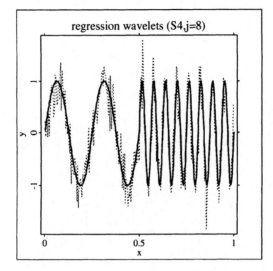

Figure 10.26: Linear wavelet estimator and true curve, $j_1 = 8$.

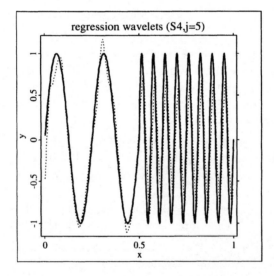

Figure 10.27: Linear wavelet estimator and true curve, with $j_1 = 5$.

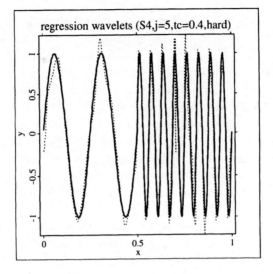

Figure 10.28: Wavelet smoother with soft threshold $0.2 \max_{jk} |\hat{\beta}_{jk}|$.

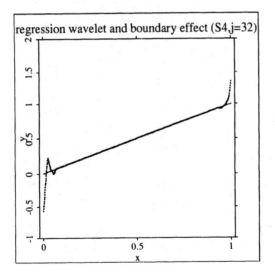

Figure 10.29: Wavelet regression with boundary effect.

Consider another example. In Figure 10.29 we plotted the function $f(x) = x$, $x \in (0,1)$ on a grid of $n = 512$ points (without observation noise) and the corresponding linear wavelet estimate $\hat{f}_{j_1}$ with $j_1 = 32$. The wavelet estimate shows well known boundary effects. A practical method for correcting the boundary problem is symmetrizing by mirroring.

We first "mirror" the original data by putting them in the reverse order symmetrically with respect to an endpoint of the interval. In the example of Figure 10.29 the mirroring with respect to $x = 1$ would result in a symmetric "tent-shaped" curve. Then we apply the usual wavelet estimation procedure with the doubled data and consider the estimator only on the original interval. Mirroring at $x = 0$ is not necessary since the symmetrized function is periodic on the doubled interval, and we use a periodically extended data for computing, cf. Chapter 12).

Figure 10.30 shows the boundary corrected estimate. The data were mirrored only at $x = 1$. The result of the wavelet estimation on this mirrored data shows that the boundary effects are no longer present.

Another important question is the choice of threshold. A variant of such a choice is to compute the following variable threshold:

$$ t = t_{jk} = \sqrt{2\hat{\sigma}_{jk}^2 \log(M_j)} \tag{10.54} $$

with

$$ \hat{\sigma}_{jk}^2 = \frac{1}{n^2} \sum_{i=1}^{n} \psi_{jk}^2(X_i) \left[\frac{2}{3} \left(Y_i - \frac{Y_{i-1} + Y_{i+1}}{2} \right)^2 \right] \tag{10.55} $$

and M_j the number of non–zero coefficients $\hat{\beta}_{jk}$ on level j. In most common cases M_j is proportional to 2^j, see Remark 10.1. The value $\hat{\sigma}_{jk}^2$ is an empirical estimator of the variance $\mathrm{Var}(\hat{\beta}_{jk})$. The term in squared brackets in the sum (10.55) is a local noise variance estimate, see Gasser, Stroka & Jennen-Steinmetz (1986). The procedure (10.54), (10.55) has been suggested by Michael Neumann. Note that the threshold (10.54) depends both on j and k. A motivation of such a threshold choice is given in Section 11.4.

10.9   Other statistical models

Besides density estimation and regression, several statistical models were studied in a wavelet framework. We mention here some of them.

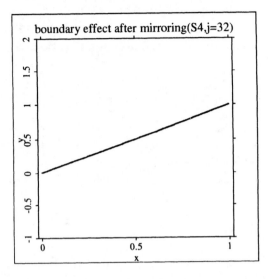

Figure 10.30: Wavelet regression estimator after mirroring.

Gaussian white noise model

This is probably the most commonly discussed model in wavelet context. It has the form of stochastic differential equation

$$dY(t) = f(t)dt + \varepsilon dW(t), \ t \in [0,1], \tag{10.56}$$

where W is the standard Brownian motion on $[0,1]$, $0 < \varepsilon < 1$, and f is an unknown function to be estimated. The observations are the values of the process $Y(t)$, $0 \le t \le 1$, satisfying (10.56).

The Gaussian white noise model was introduced by I.A. Ibragimov and R.Z.Hasminskii (see e.g. Ibragimov & Hasminskii (1981)). It appeared first as a convenient idealization of the nonparametric regression model with regular design. In particular, the analogy is established by setting $\varepsilon = 1/\sqrt{n}$, and considering asymptotics as $\varepsilon \to 0$. The model (10.56) reduces technical difficulties and is a perfect guide to more applied statistical problems. Moreover, it seems that recent works involving constructive equivalence of experiments could allow to extend this property of guiding principle to a real transfer of the results obtained in the Gaussian white noise model to more difficult settings (see for instance Brown & Low (1996), Nussbaum (1996)).

To define wavelet estimators in this model one has to use the same formulae as before in the chapter, with the only modification: $\hat{\alpha}_{jk}$ and $\hat{\beta}_{jk}$ should be of the form

$$\hat{\alpha}_{jk} = \int \varphi_{jk}(t) \, dY(t), \; \hat{\beta}_{jk} = \int \psi_{jk}(t) dY(t). \tag{10.57}$$

Clearly, these stochastic integrals are unbiased estimators of α_{jk} and β_{jk} under the model (10.56). For a detailed discussion of wavelet thresholding in this model see Donoho, Johnstone, Kerkyacharian & Picard, (1995, 1997).

Time series models

Gao(1993b, 1993a),Moulin (1993) investigated the behavior of wavelet estimates in time series analysis. Neumann(1996a, 1996b) has put the thresholding results into a unified approach permitting to treat a lot of different models. Neumann & von Sachs (1995) give a brief overview on wavelet thresholding in non-Gaussian and non-iid situations, respectively. They establish joint asymptotic normality of the empirical coefficients and apply non-linear adaptive shrinking schemes to estimate the spectral density.

Recently, there has been growing interest in wavelet estimation of the dependence structure of non stationary processes with locally stationary or "slowly varying" behavior. See for example Dahlhaus (1997), von Sachs & Schneider (1996), Neumann & von Sachs (1997), Donoho, Mallat & von Sachs (1996).

Diffusion models

Genon-Catalot, Laredo & Picard (1992) described the behavior of a linear wavelet estimator of a time varying diffusion coefficient observed at discrete times. Hoffmann (1996) provided the non linear wavelet estimator of a time or state varying diffusion coefficient, observed at discrete times. He showed that this estimator attains optimal rates of convergence on a large scale of smoothness classes.

Images

It is possible to generalize the wavelet tools to the multivariate case. A multivariate extension of MRA was introduced by Mallat (1989). Nason & Silverman (1994), Ogden (1997) give details how to compute the corresponding wavelet estimators in the case of two-dimensional images.

Some work has been done on the wavelet estimators based on the product of d univariate wavelet bases (Tribouley (1995), Delyon & Juditsky (1996a), Neumann & von Sachs (1995), Neumann(1996a, 1996b)).

Tribouley (1995) showed that the wavelet thresholding procedure, under a certain threshold choice, attains optimal rates of convergence on the multivariate Besov classes for the density estimation problem. Delyon & Juditsky (1996a) generalized these results and considered the nonparametric regression setting as well. In these papers only isotropic multivariate Besov classes were studied, i.e. the case where the smoothness of estimated function is the same in all directions. Neumann & von Sachs (1995) and Neumann (1996a, 1996b) showed that the product wavelet estimators can attain minimax rates of convergence in anisotropic smoothness classes.

A quite natural application of this methodology can be found in Neumann & von Sachs (1995) to the particular problem of estimating the time-varying spectral density of a locally stationary process. In this case the two axes on the plane, time and frequency, have a specific meaning. Accordingly, one cannot expect the same degrees of smoothness in both directions. Hence, the use of the anisotropic basis seems to be more natural than the use of the isotropic one.

Chapter 11

Wavelet thresholding and adaptation

11.1   Introduction

This chapter treats in more detail the adaptivity property of nonlinear (thresholded) wavelet estimates. We first introduce different modifications and generalizations of soft and hard thresholding. Then we develop the notion of adaptive estimators and present the results about adaptivity of wavelet thresholding for density estimation problems. Finally, we consider the data–driven methods of selecting the wavelet basis, the threshold value and the initial resolution level, based on Stein's principle. We finish by a discussion of oracle inequalities and miscellaneous related topics.

11.2   Different forms of wavelet thresholding

Two simplest methods of wavelet thresholding (soft and hard thresholding) were introduced already in Chapter 10. Here we give a more detailed overview and classification of the available thresholding techniques. For definiteness, we assume that the problem of density estimation is considered. Thus, we have a sample $X_1, \ldots, X_n$ of n i.i.d. observations from an unknown density f, and we want to estimate f. Extension of the definitions given below to other models (nonparametric regression, Gaussian white noise model, spectral density estimation etc.) is standard, and it can be established in the same spirit as discussed in Chapter 10. We classify the thresholding procedures into three groups: local, global and block thresholding. For local thresholding we distinguish between fixed and variable thresholding techniques.

Local thresholding

These are essentially the procedures of the type of soft and hard thresholding introduced in Chapter 10. The word "local" means that individual coefficients independently of each other are subject to a possible thresholding.

Let $\hat{\beta}_{jk}$ be the empirical wavelet coefficients defined in (10.3), and let $\eta_{jk}(u)$ be a function of $u \in \mathbb{R}$. It is possible that η_{jk} is a random function depending on $X_1, \ldots, X_n$. Assume that

$$\eta_{jk}(u) = 0, \quad |u| \le t,$$

where $t > 0$ is a threshold (possibly random). The local thresholded empirical wavelet coefficients are

$$\beta_{jk}^* = \eta_{jk}(\hat{\beta}_{jk}). \tag{11.1}$$

For example, in the soft and hard thresholding defined in Chapter 10 the functions η_{jk} are non–random, do not depend on j, k, and have the form, respectively

$$\eta_{jk}(u) = \eta^S(u) = (|u| - t)_+ \text{sign } u \tag{11.2}$$

$$\eta_{jk}(u) = \eta^H(u) = u \, I\{|u| > t\}. \tag{11.3}$$

The wavelet density estimator with the coefficients (11.1) has the form

$$f^*(x) = \sum_k \hat{\alpha}_{j_0 k} \varphi_{j_0 k}(x) + \sum_{j=j_0}^{j_1} \sum_k \eta_{jk}(\hat{\beta}_{jk}) \psi_{jk}(x). \tag{11.4}$$

We call it *local thresholding wavelet estimator*. It follows from Proposition 10.3 that the choice of threshold

$$t = c\sqrt{\frac{\log n}{n}}, \tag{11.5}$$

where $c > 0$ is a suitably chosen constant, guarantees the asymptotically optimal (up to a log–factor) behavior of f^* when $\eta_{jk}(u) = \eta^H(u)$. A similar result is true for the case of soft thresholding. The question how to choose c is not answered by these results (we know only that c should be large enough).

Other types of thresholding, where η_{jk} depends on j (and not on k), are defined by (11.2) and (11.3) with

$$t = t_j = c\sqrt{\frac{j - j_0}{n}} \tag{11.6}$$

(Delyon & Juditsky (1996a)), or with

$$t = t_j = c\sqrt{\frac{j}{n}} \tag{11.7}$$

(Tribouley (1995),Donoho, Johnstone, Kerkyacharian & Picard (1996)). Here again $c > 0$ is a suitable constant.

Finally, the example of η_{jk} depending on both j and k is provided by the soft thresholding (11.2) or (11.3) with

$$t = t_{jk} = \sqrt{2\sigma_{jk}^2[\psi] \log M_j}, \tag{11.8}$$

where $\sigma_{jk}^2[\psi]$ is the variance of the empirical wavelet coefficient $\hat{\beta}_{jk}$ and M_j is the number of non–zero coefficients on level j. We shall discuss the threshold choice (11.8) later in this chapter. As $\sigma_{jk}^2[\psi]$ is not known, one should replace it by its empirical version. This leads to a random threshold $t = t_{jk}$ (respectively random function η_{jk}).

If the threshold t of the local thresholding estimator is the same for all j, k (as in (11.5)), we call f^* the estimator with *fixed* threshold. Otherwise, if t may vary with j and/ or k (as in (11.6)-(11.8)), f^* is called *local thresholding wavelet estimator with variable threshold*.

Global thresholding

Instead of keeping or deleting individual wavelet coefficients, one can also keep or delete a whole j-level of coefficients. This leads to the following definition of the wavelet estimator:

$$f^*(x) = \sum_k \hat{\alpha}_{j_0 k}\varphi_{j_0 k}(x) + \sum_{j=j_0}^{j_1} \eta_j\left(\sum_k \hat{\beta}_{jk}\psi_{jk}(x)\right) \tag{11.9}$$

where $\eta_j(\cdot)$ is some non-linear thresholding type transformation. Kerkyacharian, Picard & Tribouley (1996) considered such an estimator of a probability density f. They proposed the following analogues of hard and soft thresholding respectively:

$$\eta_j^H(u) \;=\; uI\left\{S_j(p) > \frac{2^j}{n^{p/2}}\right\}, \tag{11.10}$$

$$\eta_j^S(u) \;=\; u\left(\frac{S_j(p) - \frac{2^j}{n^{p/2}}}{S_j(p)}\right)_{+}, \tag{11.11}$$

where $S_j(p)$ is a certain statistic depending on $X_1, \ldots, X_n$ and $p \geq 1$ is a parameter. In particular, if p is an even integer, $p \leq n$, $S_j(p)$ is defined as

$$S_j(p) = \frac{1}{\binom{n}{p}} \sum_{i_1 \neq \ldots \neq i_p} \sum_k \psi_{jk}(X_{i_1}) \cdots \psi_{jk}(X_{i_p}).$$

The definition of $S_j(p)$ for general p is given in Kerkyacharian et al. (1996). The estimator f^* defined in (11.9), with $\eta_j = \eta_j^H$ or $\eta_j = \eta_j^S$, is called *global thresholding wavelet density estimator*. We discuss later the advantages and drawbacks of this estimate. Let us now make only some general remarks:

- The above definition of global thresholding estimator is completely data–driven, which is not the case for local thresholding estimators with the threshold values (11.5)–(11.7).

- The computational aspects become more difficult when p increases. The constant p, as we shall see later, comes from the L_p loss function that we want to optimize.

- This procedure provides a L_p-generalization of a method introduced in the L_2-setting and the context of Fourier series by Efroimovich (1985). The expression (11.11) is reminiscent of the James-Stein estimator, see Ibragimov & Hasminskii (1981), Chapter 1. It is also close to a procedure introduced by Lepskii (1990) in the context of kernel estimates.

Block thresholding

Block thresholding is a procedure intermediate between local and global thresholding. It keeps or deletes specially chosen blocks of wavelet coefficients on each level. Such a method was introduced by Hall, Kerkyacharian & Picard(1996a, 1996c). It is defined as follows. Divide the set of all integers into non–overlapping blocks of length $l = l(n)$:

$$B_k = \{m : (k-1)l + 1 \ \leq m \ \leq kl\}, \ k \in \mathbb{Z}.$$

Put

$$b_{jk} \ = \ \frac{1}{l} \sum_{m \in B_k} \beta_{jm}^2.$$

Take the following estimator of b_{jk}:

$$\hat{b}_{jk} \ = \ \frac{1}{l} \sum_{m \in B_k} \hat{\beta}_{jm}^2,$$

and define the wavelet estimator of a density f as:

$$f^*(x) = \sum_k \hat{\alpha}_{j_0 k} \varphi_{j_0 k}(x) + \sum_{j=j_0}^{j_1} \sum_k \left(\sum_{m \in B_k} \hat{\beta}_{jk} \psi_{jm}(x) \right) I\left\{ \hat{b}_{jk} > c n^{-1} \right\}, \quad (11.12)$$

where $c > 0$ is a constant controlling the threshold. This estimate f^* is called *block thresholding wavelet density estimator.*

In most cases, the block estimator has better asymptotic properties than the local thresholding estimators, since it has no additional logarithmic factor in the rate of convergence (see Hall, Kerkyacharian & Picard(1996a, 1996c) for the details).

An obvious drawback of the estimator (11.12), as compared to the global thresholding estimator (11.9)–(11.11), is again the fact that it is not completely data–driven. It depends on the constant c which is not given explicitly by the theory, and has to be chosen in some empirical way (this constant is given by the theory up to the knowledge of the uniform bound of f, see Chapter 10).

11.3   Adaptivity properties of wavelet estimates

The wavelet estimators defined above and in Chapter 10 require prior knowledge of several parameters:

1) the highest level j_1 and the initial level j_0,

2) the threshold t, or more generally, the vector of thresholds $\mathbf{t} = \{t_{jk}\}_{j,k}$,

3) the wavelet basis $\{\varphi_{jk}, \psi_{jk}\}$, or, equivalently, the father wavelet φ (under the assumption that mother wavelet ψ is related to φ by a fixed transformation, to avoid non-uniqueness cf. Section 5.2).

In Chapter 10 we specified some assumptions on these parameters that guarantee near optimal asymptotic behavior of wavelet estimates. These assumptions are formulated in terms of the regularity m (or s) of the estimated function. In practice this is a serious drawback since, in general, it is impossible to know the regularity of the functional class where the function sits. Moreover, a single function may be in the intersection of different classes. For instance, consider the following example of a "2–bumps" function g. Assume that g coincides with $|x|$ on $[-1/2, 1/2]$, is extremely regular outside this interval and compactly supported. Its derivative satisfies

$$g'(x) = -I\{x \in [-1/2, 0]\} + I\{x \in [0, 1/2]\}$$

on $[-1/2, 1/2]$ and g' is a very regular function outside $[-1/2, 1/2]$. If we look at $\|\tau_h g' - g'\|_p$ it is, clearly, of order $(2h)^{1/p}$. Hence, $g' \in B_p^{1/p,\infty}$ for every $1 \le p < \infty$. We conclude that g belongs to all the spaces $B_p^{1+1/p,\infty}$, $1 \le p < \infty$.

Another example is given by the function

$$f(x) = \sum_{k=1}^{2^j} 2^{-\frac{3j}{2}} \psi_{jk}(x)$$

where ψ is a mother wavelet of a MRA: clearly f belongs to all the spaces $B_p^{1,1}$, $\forall p \ge 1$. The results of Chapter 10 entail that different spaces are characterized by different optimal convergence rates of estimators. Thus, it is important to find an estimator attaining simultaneously the best rates of convergence on a large scale of spaces (respectively, functional classes). Fortunately, wavelet estimators enjoy this property.

Let A be a given set and let $\{\mathcal{F}_\alpha, \alpha \in A\}$ be the scale of functional classes $\mathcal{F}_\alpha$ indexed by $\alpha \in A$. (For example, $\alpha \in [0, 1]$, $\mathcal{F}_\alpha$ is a unit ball in $B^{\alpha\infty}_\infty$.) Denote by $R_n(\alpha, p)$ the minimax risk over $\mathcal{F}_\alpha$ for the L_p-loss:

$$R_n(\alpha, p) = \inf_{\hat{f}} \sup_{f \in \mathcal{F}_\alpha} E_f \|\hat{f} - f\|_p^p.$$

DEFINITION 11.1 *The estimator f^* is called* **adaptive** *for L_p-loss and the scale of classes $\{\mathcal{F}_\alpha, \alpha \in A\}$ if for any $\alpha \in A$ there exists $c_\alpha > 0$ such that*

$$\sup_{f \in \mathcal{F}_\alpha} E_f \|f^* - f\|_p^p \leq c_\alpha R_n(\alpha, p), \quad \forall n \geq 1.$$

The estimator f^* is called **adaptive up to a logarithmic factor** *for L_p-loss and the scale of classes $\{\mathcal{F}_\alpha, \alpha \in A\}$ if for any $\alpha \in A$ there exist $c_\alpha > 0$ and $\gamma = \gamma_\alpha > 0$ such that*

$$\sup_{f \in \mathcal{F}_\alpha} E_f \|f^* - f\|_p^p \leq c_\alpha (\log n)^\gamma R_n(\alpha, p), \quad \forall n \geq 1.$$

Thus, as far as the rate of convergence is concerned, the adaptive estimator is optimal and behaves itself as if it knows in advance in which class the function lies (i.e. as if it knows α). For more insight into the general problem of adaptivity we refer to Lepskii(1990, 1991, 1992), Lepski & Spokoiny (1995), Lepski, Mammen & Spokoiny (1997), Birgé & Massart (1997).

Below we present without proof some results illustrating that the wavelet estimators have the above adaptation property. Let us take again the density estimation framework.

In the following two propositions we assume that $\mathcal{F}_\alpha$ is a Besov class: $\mathcal{F}_\alpha = \tilde{B}(s, r, q, L)$, where $\alpha = (s, r, q, L)$

$$\tilde{B}(s, r, q, L) = \{f : f \text{ is a probability density on } \mathbb{R} \text{ with a compact support of length} \leq L', \text{ and } \|f\|_{srq} \leq L\}.$$

Here s, r, p, q, L, L' are positive numbers. The knowledge of the parameter L' is not necessary for the construction of the estimates. Therefore we do not include it into α.

PROPOSITION 11.1 *(Donoho, Johnstone, Kerkyacharian & Picard (1996))
Let the father wavelet φ satisfy the conditions of Theorem 9.4 for some integer $N > 0$. Let L be a given positive number. The local thresholding estimate
chosen so that $j_0 = 0$, $2^{j_1} \simeq \frac{n}{\log n}$, $t = c\sqrt{\frac{\log n}{n}}$, (where c is a constant depending on L), is adaptive up to a logarithmic factor for any loss L_p, $1 \leq p < \infty$,
and the scale of classes $\{\mathcal{F}_\alpha, \ \alpha \in A\}$ where*

$$A = (1/r, N) \times [1, \infty] \times [1, \infty] \times \{L\}.$$

Recall that N here is the number of vanishing moments of the mother
wavelet ψ (see Chapters 9 and 10).

PROPOSITION 11.2 *(Kerkyacharian et al. (1996)) Let the father wavelet
φ satisfy the conditions of Theorem 9.4 for some integer $N > 0$. Let $r \geq 1$
be a given number. The global thresholding estimate defined with (11.10),
(11.11), where $p = r$, and such that $j_0 = 0$, $2^{j_1} \simeq \frac{n}{\log n}$, is adaptive for any
loss L_p, $1 \leq p \leq r$, and the scale of classes $\{\mathcal{F}_\alpha, \ \alpha \in A\}$ where*

$$A = (1/r, N) \times \{r\} \times [1, \infty] \times (0, \infty).$$

We stated the two propositions together to simplify the comparison. The
propositions deal with the local and global procedures respectively. As it
can be seen, the limitations with respect to the regularity s are the same
for both procedures: $s \in (1/r, N)$. The local procedure always looses a
logarithmic factor, but its range of loss functions is wider. The range of r is
very limited in the case of global thresholding (r should be known), whereas
there is no limitation in the local estimate. It is precisely this fact which
is described by saying that local thresholding estimate is able to adapt to
"inhomogeneous irregularities". Finally, the adaptation with respect to the
radius L of the Besov ball is very poor in the local case: L should be known.
This is essentially because the constant c depends on L.

REMARK 11.1 For the global thresholding estimate, the result of Proposition 11.1 have been generalized to the case of dependent data with β-mixing
conditions by Tribouley & Viennet (1998). For the local estimate, the adaptation property of Proposition 11.1 has been obtained in a number of very
different situations. Among others let us cite Donoho, Johnstone, Kerkyacharian & Picard (1995), concerning the Gaussian white noise model and

regression, Johnstone & Silverman (1997) concerning regression with dependent data, Wang (1996), Neumann & von Sachs (1997), Hoffmann (1996), concerning the time series models. Similar results can be obtained in inverse problems using the "wavelet-vaguelette" decomposition of Donoho (1995).

REMARK 11.2 In the same spirit, let us also summarize the performance of the block thresholding estimate. By choosing

$$2^{j_0} \simeq n^{1/(1+2N)}, (\text{ where } N \text{ is the number of zero moments of } \psi),$$

$$2^{j_1} \simeq \frac{n}{\log n}, \quad l(n) \simeq (\log n)^2,$$

with c depending on L, we obtain adaptivity for the L_2-loss, without any additional logarithmic factor, when α is in the range

$$\alpha \in (1/2, N) \times \{2\} \times [1, \infty] \times \{L\}.$$

This holds for a much wider class $\mathcal{F}_\alpha$ than above. Here $\mathcal{F}_\alpha$ can be the set of densities f with compact support, $f = f_1 + f_2$, where f_1 is a "regular" function, $\|f_1\|_{srq} \leq L$, and f_2 is a "perturbation": a bounded function containing irregularities such as discontinuities, Doppler or Chirps oscillations (see Hall, Kerkyacharian & Picard (1996c))

11.4 Thresholding in sequence space

In studying the properties of wavelet estimates it is often useful to introduce an idealized statistical model (called *sequence space model*), that approximates the true one.

Let $\hat{\alpha}_{j_0 k}, \hat{\beta}_{jk}$ be the empirical wavelet coefficients, as defined in Section 10.2. Clearly, one can write

$$\begin{aligned} \hat{\alpha}_{j_0 k} &= \alpha_{j_0 k} + \sigma_{j_0 k}[\varphi]\zeta_{j_0 k}, \\ \hat{\beta}_{jk} &= \beta_{jk} + \sigma_{jk}[\psi]\xi_{jk}, \end{aligned} \qquad (11.13)$$

where $\alpha_{j_0 k}, \beta_{jk}$ are the "true" wavelet coefficients, $\zeta_{j_0 k}, \xi_{jk}$ are random variables with zero mean and variance 1, and $\sigma_{j_0 k}[\varphi], \sigma_{jk}[\psi]$ are the corresponding scale factors. (Note that $E(\zeta_{j_0 k}) = 0$, $E(\xi_{jk}) = 0$, since $\hat{\alpha}_{j_0 k}$ and $\hat{\beta}_{jk}$ are unbiased estimators of $\alpha_{j_0 k}$ and β_{jk} respectively.)

Since the standard thresholding procedures are applied only to $\hat{\beta}_{jk}$ coefficients ("detail coefficients") we discuss the approximation in sequence space model for $\hat{\beta}_{jk}$ on a fixed level j.

We assume here and below that we deal with compactly supported wavelets φ and ψ. Therefore, only a finite number M of wavelet coefficients $\hat{\beta}_{jk}$ is non-zero, and we can assume that k varies from 1 to M. Also, note that ξ_{jk} are asymptotically Gaussian (since $\hat{\beta}_{jk}$ is a sum of independent random variables), and ξ_{jk} is approximately noncorrelated with $\xi_{jk'}, k \neq k'$. In fact, if ψ is compactly supported, $supp\ \psi \subseteq [-A, A]$, for some $A > 0$, then

$$\int \psi_{jk}(x)\psi_{jk'}(x)f(x)dx = 0, \tag{11.14}$$

whenever $|k - k'| > 2A$. Hence, in the case $|k - k'| > 2A$ the covariance

$$\begin{aligned}
\mathrm{Cov}(\hat{\beta}_{jk}, \hat{\beta}_{jk'}) &= E\left(\frac{1}{n^2}\sum_{i,m=1}^{n}\psi_{jk}(X_i)\psi_{jk'}(X_m)\right) - E(\hat{\beta}_{jk})E(\hat{\beta}_{jk'}) \\
&= \frac{1}{n^2}\sum_{i=1}^{n}E\left(\psi_{jk}(X_i)\psi_{jk'}(X_i)\right) - \frac{1}{n}E(\hat{\beta}_{jk})E(\hat{\beta}_{jk'}) \\
&= \frac{1}{n}\int \psi_{jk}(x)\psi_{jk'}(x)f(x)dx - \frac{1}{n}\beta_{jk}\beta_{jk'} \\
&= -\frac{1}{n}\beta_{jk}\beta_{jk'},
\end{aligned}$$

and since $\beta_{jk} = O(2^{-j/2})$, the covariance for j large enough is much smaller than the variance

$$\begin{aligned}
\sigma_{jk}^2[\psi] = Var(\hat{\beta}_{jk}) &= \frac{1}{n}\left[E\left(\psi_{jk}^2(X_1)\right) - E^2\left(\psi_{jk}(X_1)\right)\right] \tag{11.15} \\
&= \frac{1}{n}\left[E\left(\psi_{jk}^2(X_1)\right) - \beta_{jk}^2\right] = O\left(\frac{1}{n}\right),
\end{aligned}$$

as $n \to \infty$.

This suggests that, in a certain asymptotical approximation (which we do not pretend to develop here with full mathematical rigour), the "new" observation model (11.13) is equivalent to the *sequence space model*:

$$Z_k = \theta_k + \sigma_k\xi_k, \quad k = 1, \ldots, M, \tag{11.16}$$

where Z_k plays the role of $\hat{\beta}_{jk}$, while θ_k is an unknown parameter (it stands for the true coefficient β_{jk}). Here ξ_k are i.i.d. $\mathcal{N}(0,1)$ random variables and $\sigma_k > 0$. Let us remark once again that (11.16) is an idealized model for wavelet coefficients of a *fixed level j*. We drop the index j as compared to (11.13) since the level j is fixed. The integer M in (11.16) is arbitrary, but one may think that $M \sim 2^j$ to translate the argument back into the wavelet context.

In the sequence space model (11.16) our aim is to estimate the unknown vector of parameters

$$\theta = (\theta_1, \ldots, \theta_M),$$

given the vector of Gaussian observations $\mathbf{z} = (Z_1, \ldots, Z_M)$. The sequence space model (11.16) can be used as an approximation for the study of nonparametric wavelet estimators in other models for example in regression and Gaussian white noise models.

Note that in the Gaussian white noise case (see (10.56),(10.57)) the errors ξ_{jk} in (11.13) are i.i.d. Gaussian $\mathcal{N}(0,1)$ random variables and $\sigma_{jk}[\psi] = \varepsilon$. Thus, the corresponding sequence space model is

$$Z_k = \theta_k + \varepsilon \xi_k, \ \xi_k \sim \mathcal{N}(0,1).$$

In this case the sequence space model is exactly (and not only approximately) equivalent to the original model.

Sequence space models allow to provide a reasonable interpretation of some threshold rules introduced earlier in this chapter. Let us first analyse the Gaussian white noise case. It is well known (see e.g. Leadbetter, Lindgren & Rootzén (1986)) that for M i.i.d. standard Gaussian variables $\xi_1, \ldots, \xi_M$ one has $P\left(\max_{1 \leq k \leq M} |\xi_k| \geq \sqrt{2 \log M} \right) \to 0$, as $M \to \infty$. Therefore if the threshold is set to $t = \varepsilon \sqrt{2 \log M}$, a pure noise signal (i.e. $\theta_1 = \ldots = \theta_M = 0$) is with high probability correctly estimated as being identically zero: it makes no sense to increase t above $\varepsilon \sqrt{2 \log M}$. Note that, as M is proportional to 2^j, the threshold t is in fact of the form $c\varepsilon \sqrt{j}$ for some constant $c > 0$.

The choice $t = \varepsilon \sqrt{2 \log n}$ where n is the total number of observations, allows to estimate correctly the zero signal for all coefficient levels j (in fact, $n > M$). This threshold choice, called *universal threshold*, typically kills most of the coefficients and leaves only few large coefficients intact. As a result, visually the picture of the wavelet estimator looks smooth: no small spikes are present. This is achieved on the expense of a loss in the precision of estimation as compared to more sophisticated thresholding techniques.

Let us turn now to the general sequence space model (11.16). Quite a similar reasoning gives the variable thresholds $t_k = \sigma_k\sqrt{2\log M}$ for different coefficients θ_k. As $\sigma_k \sim \frac{1}{\sqrt{n}}$ in the density estimation case (see (11.15)), this yields $t_k = c_k\sqrt{\frac{j}{n}}$ where $c_k > 0$ is a constant depending on k. This explains the variable thresholding procedures (11.7) and (11.8) as well as their empirical counterparts (see (10.54), (10.55)) and Remark 11.3 below). The fixed threshold choice $t = c\sqrt{\frac{\log n}{n}}$ is motivated by analgous considerations, since the number of levels j kept in the wavelet estimator is typically of $O(\log n)$ order (see Sections 10.2,10.4).

The universal threshold can be defined for general sequence space model (11.16) as well: Donoho & Johnstone (1995) introduce it in the form

$$t = \hat{\sigma}\sqrt{\frac{2\log n}{n}},$$

where $\hat{\sigma}$ is the robust estimate of scale defined as the median absolute deviation (MAD) of the empirical wavelet coefficients corresponding to the highest resolution level j_1. The reason for using only the highest level coefficients for the purpose of variance estimation is that they consist mostly of noise, in contrast to the lower level coefficients that are believed to contain information on the significant features of the estimated function. The MAD universal thresholding estimator is simple and often used in practice. Observe that the universal thresholding tends to oversmooth the data, as already mentioned above.

A number of heuristic thresholding techniques is based on parametric hypothesis testing for the Gaussian sequence space model framework. A recent proposal by Abramovich & Benjamini (1996) is designed to control the expected proportion of incorrectly included coefficients among those chosen for the wavelet reconstruction. The objective of their procedure is to include as many coefficients as possible provided that the above expected proportion is kept below a given value. A tendency to increase the number of coefficients, in general, leads to undersmoothing. However, if the estimated function has several abrupt changes this approach appears to be useful. The corresponding simulation study can be found in Abramovich & Benjamini (1996). A different testing procedure is proposed by Ogden & Parzen (1996). They perform a levelwise rather than overall testing. At each level, they test the null hypothesis of a pure Gaussian noise signal ($\theta_1 = \ldots = \theta_M = 0$). If this hypothesis is rejected (i.e. if a significant signal is present) the largest coefficient in absolute value is kept aside, and then the test is repeated with the remaining coefficients. Iterating this procedure, one finally arrives, at each resolution level, to a classification of the coefficients into two groups: large coefficients that are believed to contain some information on the signal, and small coefficients statistically indistinguishable from the pure noise. Finally, only the large coefficients are included in the wavelet estimator. This gives us an example of local variable thresholding with random mechanism. Juditsky (1997) developped a different but somewhat related thresholding approach, applying the implicit bias – variance comparison procedure of Lepskii (1990). This method, again, is charaterized by a random local variable thresholding. The idea of the method is formulated for the sequence space model and extended to the equispaced design regression and density estimation problems. Juditsky (1997) proves that for these problems his wavelet estimator is adaptive for the L_p-losses on the scale of Besov classes in the sense of Definition 11.1.

11.5 Adaptive thresholding and Stein's principle

In this section we discuss the data driven choice of threshold, initial level j_0 and the wavelet basis by the Stein (1981) method of unbiased risk estimation. The argument below follows Donoho & Johnstone (1995).

We first explain the Stein method for the idealized one-level observation model discussed in the previous section:

$$Z_k = \theta_k + \sigma_k \xi_k, \quad k = 1, \ldots, M, \tag{11.17}$$

where $\theta = (\theta_1, \ldots, \theta_M)$ is the vector of unknown parameters, $\sigma_k > 0$ are known scale parameters and ξ_k are i.i.d. $\mathcal{N}(0, 1)$ random variables.

Let $\hat{\theta} = (\hat{\theta}_1, \ldots, \hat{\theta}_M)$ be an estimator of θ. Introduce the mean squared risk of $\hat{\theta}$:

$$\mathcal{R} = \sum_{k=1}^{M} E(\hat{\theta}_k - \theta_k)^2.$$

Assume that the estimators $\hat{\theta}_k$ have the form

$$\hat{\theta}_k = Z_k + H_t(Z_k), \qquad (11.18)$$

where t is a parameter and $H_t(\cdot)$ is a weakly differentiable real valued function for any fixed t. One may think initially of t to be a threshold (see the example (11.21) later in this section), but Stein's argument works in the general case as well. The parameter t can be chosen by the statistician. In other words, (11.18) defines a family of estimators, indexed by t, and the question is how to choose an "optimal" $t = t^*$. Define the optimal t^* as a minimizer of the risk $\mathcal{R}$ with respect to t.

If the true parameters θ_k were known, one could compute t^* explicitly. In practice this is not possible, and one chooses a certain approximation $\hat{t}$ of t^* as a minimizer of an unbiased estimator $\hat{\mathcal{R}}$ of the risk $\mathcal{R}$. To construct $\hat{\mathcal{R}}$, note that

$$E(\hat{\theta}_k - \theta_k)^2 = E(R(\sigma_k, Z_k, t)), \qquad (11.19)$$

where

$$R(\sigma, x, t) = \sigma^2 + 2\sigma^2 \frac{d}{dx} H_t(x) + H_t^2(x).$$

In fact,

$$E(\hat{\theta}_k - \theta_k)^2 = \sigma_k^2 + 2\sigma_k E(\xi_k H_t(Z_k)) + E(H_t^2(Z_k)),$$

and, by partial integration,

$$
\begin{aligned}
E(\xi_k H_t(\theta_k + \sigma_k \xi_k)) &= \frac{1}{\sqrt{2\pi}} \int \xi H_t(\theta_k + \sigma_k \xi) e^{-\frac{\xi^2}{2}} d\xi \\
&= \frac{1}{\sqrt{2\pi}} \int H_t(\eta) \frac{(\eta - \theta_k)}{\sigma_k} \exp\left(-\frac{(\eta - \theta_k)^2}{2\sigma_k^2}\right) d\eta \\
&= \frac{1}{\sqrt{2\pi}} \int \exp\left(-\frac{(\eta - \theta_k)^2}{2\sigma_k^2}\right) \frac{dH_t(\eta)}{d\eta} d\eta \\
&= \sigma_k E\left(\frac{dH_t(x)}{dx}\Big|_{x=Z_k}\right).
\end{aligned}
$$

Thus (11.19) follows.

The relation (11.19) yields $\mathcal{R} = E(\hat{\mathcal{R}})$, where the value $\hat{\mathcal{R}} = \sum_{k=1}^{M} R(\sigma_k, Z_k, t)$ is an unbiased risk estimator, or risk predictor. It is called *Stein's unbiased risk estimator (SURE):*

$$SURE = \sum_{k=1}^{M} R(\sigma_k, Z_k, t)$$

The *Stein principle* is to minimize $\hat{\mathcal{R}}$ with respect to t and take the minimizer

$$\hat{t} = \arg\min_{t \geq 0} \sum_{k=1}^{M} R(\sigma_k, Z_k, t). \tag{11.20}$$

as a data driven estimator of the optimal t^*. The unbiasedness relation $E(\hat{\mathcal{R}}) = \mathcal{R}$ (for every t) alone does not guarantee that $\hat{t}$ is close to t^*. Some more developed argument is used to prove this (Donoho & Johnstone (1991)).

In the rest of this section we formulate the Stein principle for the example of soft thresholding wavelet estimators.

For soft thresholding (10.13) we have

$$H_t(x) = -xI\{|x| < t\} - tI\{|x| \geq t\}\text{sign}(x), \tag{11.21}$$

and

$$\begin{aligned} R(\sigma, x, t) &= (x^2 - \sigma^2)I\{|x| < t\} + (\sigma^2 + t^2)I\{|x| \geq t\} \\ &= [x^2 - \sigma^2] + (2\sigma^2 - x^2 + t^2)I\{|x| \geq t\}. \end{aligned} \tag{11.22}$$

An equivalent expression is

$$R(\sigma, x, t) = \min(x^2, t^2) - 2\sigma^2 I\{x^2 \leq t^2\} + \sigma^2.$$

The expression in square brackets in (11.22) does not depend on t. Thus, the definition (11.19) is equivalent to

$$\hat{t} = \arg\min_{t \geq 0} \sum_{k=1}^{M} (2\sigma_k^2 + t^2 - Z_k^2)I\{|Z_k| \geq t\}. \tag{11.23}$$

Let $(p_1, \ldots, p_M)$ be the permutation ordering the array $|Z_k|$, $k = 1, \ldots, M$: $|Z_{p_1}| \leq |Z_{p_2}| \leq, \ldots, \leq |Z_{p_M}|$, and $|Z_{p_0}| = 0$. According to (11.23) one obtains

$$\hat{t} = |Z_{p_l}|, \tag{11.24}$$

where

$$l = \arg \min_{0 \le k \le M} \sum_{s=k+1}^{M} (2\sigma_{p_s}^2 + Z_{p_k}^2 - Z_{p_s}^2). \qquad (11.25)$$

In particular for $M = 1$ the above equation yields the following estimator

$$\hat{\theta}_1 = \begin{cases} Z_1, & Z_1^2 \ge 2\sigma_1^2, \\ 0, & Z_1^2 < 2\sigma_1^2. \end{cases}$$

It is easy to see that computation of $\hat{t}$ defined in (11.24), (11.25) requires approximately $M \log M$ operations provided that quick sort algorithm is used to order the array $|Z_k|$, $k = 1, \ldots, M$.

Now we proceed from the idealized model (11.17) to a more realistic density estimation model. In the context of wavelet smoothing the principle of unbiased risk estimation gives the following possibilities for adaptation:

(i) *adaptive threshold choice at any resolution level $j \ge j_0$,*

(ii) *adaptive choice of j_0 plus (i),*

(iii) *adaptive choice of father wavelet $\varphi(\cdot)$ and mother wavelet $\psi(\cdot)$ plus (ii).*

To demonstrate these possibilities consider the family of wavelet estimators

$$f^*(x, \mathbf{t}, j_0, \varphi) = \sum_k \alpha_{j_0 k}^*[\varphi, t] \varphi_{j_0 k}(x) + \sum_{j=j_0}^{j_1} \sum_k \beta_{jk}^*[\psi, t_j] \psi_{jk}(x), \qquad (11.26)$$

where $\alpha_{j_0 k}^*[\varphi, t] = \hat{\alpha}_{j_0 k} + H_t(\hat{\alpha}_{j_0 k})$ and $\beta_{jk}^*[\psi, t] = \hat{\beta}_{jk} + H_t(\hat{\beta}_{jk})$ are soft thresholded empirical wavelet coefficients (cf. (10.2), (10.3), (10.13)) with $H_t(\cdot)$ from (11.21). Here $\mathbf{t} = (t, t_{j_0}, \ldots, t_{j_1})$ is a vector of thresholds. The dependence of f^* on ψ is skipped in the notation since the mother wavelet ψ is supposed to be canonically associated with the father wavelet (see Section 5.2). As in (11.19) it can be shown that, under certain general conditions,

$$E\|f^* - f\|_2^2 = E\left(\hat{R}(\mathbf{t}, j_0, \varphi)\right).$$

Here Stein's unbiased risk estimator is given by

$$\hat{R}(\mathbf{t}, j_0, \varphi) = \sum_k R(\sigma_{j_0 k}[\varphi], \hat{\alpha}_{j_0 k}, t) + \sum_{j=j_0}^{j_1} \sum_k R(\sigma_{jk}[\psi], \hat{\beta}_{jk}, t_j), \qquad (11.27)$$

where $R(\sigma, x, t)$ is defined in (11.22), and $\sigma_{jk}^2[\psi]$ and $\sigma_{jk}^2[\varphi]$ are variances of the corresponding empirical wavelets coefficients. To obtain the "best" estimator from the family (11.26) one can choose the unknown parameters of the estimator minimizing $\hat{\mathcal{R}}(\mathbf{t}, j_0, \varphi)$. For the cases (i),(ii),(iii) these parameters can be chosen, respectively, as follows.

(i) *Adaptive choice of thresholds:*

$$\hat{\mathbf{t}} = \arg\min_{\mathbf{t}} \hat{\mathcal{R}}(\mathbf{t}, j_0, \varphi).$$

(ii) *Adaptive choice of thresholds and j_0:*

$$(\hat{\mathbf{t}}, \hat{j}_0) = \arg\min_{\mathbf{t}, j_0} \hat{\mathcal{R}}(\mathbf{t}, j_0, \varphi).$$

(iii) *Adaptive choice of thresholds, j_0 and wavelet basis:*

$$(\hat{\mathbf{t}}, \hat{j}_0, \hat{\varphi}) = \arg\min_{\mathbf{t}, j_0, \varphi} \hat{\mathcal{R}}(\mathbf{t}, j_0, \varphi).$$

In the case (iii) it is assumed that the minimum is taken over a finite number of given wavelet bases.

Note that optimization with respect to $\mathbf{t}$ can be implemented as in the fast algorithm described in (11.24), (11.25).

REMARK 11.3 Since in practice the values $\sigma_{jk}^2[\varphi]$, $\sigma_{jk}^2[\psi]$ are not available, one can use instead their empirical versions. For example if (11.26) is the wavelet density estimator, based on the sample $X_1, \ldots, X_n$, one can replace $\sigma_{jk}^2[\psi]$ by its estimator

$$\hat{\sigma}_{jk}^2[\psi] = \frac{1}{n}\left(\frac{1}{n}\sum_{i=1}^{n} \psi_{jk}^2(X_i) - \hat{\beta}_{jk}^2\right). \tag{11.28}$$

In fact, for $\hat{\beta}_{jk}$ defined in (10.3), we have

$$
\begin{aligned}
\sigma_{jk}^2[\psi] &= \mathrm{Var}(\hat{\beta}_{jk}) \\
&= \frac{1}{n}\left(E\left(\psi_{jk}^2(X_1)\right) - \beta_{jk}^2\right).
\end{aligned}
$$

It is clear that (11.28) yields a consistent estimator of $\sigma_{jk}^2[\psi]$ under rather general assumptions on ψ_{jk} and on the underlying density of X_i's.

REMARK 11.4 If one wants to threshold only the coefficients β_{jk}, which is usually the case, the function $H_t(\cdot)$ for α_{jk} should be identically zero. Therefore, $R(\sigma_{j_0k}[\varphi], \hat{\alpha}_{jk}, t)$ in (11.26) should be replaced by $\sigma_{j_0k}[\varphi]$ and SURE takes the form

$$\hat{\mathcal{R}}\left((t_{j_0}, \ldots, t_{j_1}), j_0, \varphi\right) = \sum_k \sigma_{jk}^2[\varphi] + \sum_{j=j_0}^{j_1} \sum_k R\left(\sigma_{jk}[\psi], \hat{\beta}_{jk}, t_j\right).$$

Let us now apply the Stein principle to a regression estimation example. We choose a step function similar to our densities of Section 10.2:

$$f(x) = 0.1I(x < 0.4) + 2I(x \in [0.4, 0.6] + 0.5I(x \in [0.6, 0.8]), x \in [0, 1].$$

The function was observed at 128 equispaced points and disturbed with Gaussian noise with variance $1/128$. We use the Stein rule only for threshold choice (i) (level by level) and not for the cases (ii) and (iii) where the adaptive choice of j_0 and of the basis is considered. We thus choose the threshold $\hat{\mathbf{t}}$ as the minimizer with respect to $\mathbf{t} = (t_{j_0}, \ldots, t_{j_1})$ of

$$\hat{\mathcal{R}}(\mathbf{t}) = \sum_{j=j_0}^{j_1} \sum_k R\left(\hat{\sigma}_{jk}[\psi], \hat{\beta}_{jk}, t_j\right)$$

where as above $R(\sigma, x, t)$ is defined in (11.22) and $\hat{\sigma}_{jk}[\psi]$ is an empirical estimator of the variance of the wavelet regression coefficients. For computation we use the discrete wavelet transform based methods described in Chapter 12 below.

In Figure 11.1 we display the true regression function together with the noisy data. The next Figure 11.2 presents the result of *SURE* estimation. The true curve is shown in both plots as a dashed line.

11.6   Oracle inequalities

Instead of taking the minimax point of view, to describe the performance of estimators, one can also provide concise accounts of mean squared error for single functions. This is precisely discussed in the papers of Hall & Patil (1995a, 1995b). This approach shows particularly that the local thresholding does not achieve an effective balance of bias against variance at a first-order

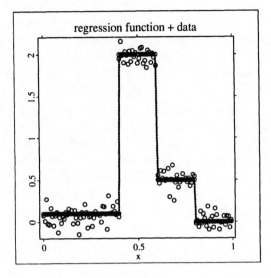

Figure 11.1: Regression function and the noisy data.

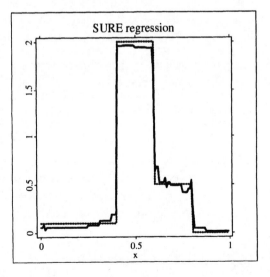

Figure 11.2: SURE regression estimator and the regression function.

level. Such a balance may be achieved by suitable adjusting of the primary resolution level, but then the price to pay is adaptivity. In contrast, the block thresholding rules permit this balance between bias and variance and preserve adaptivity (see Hall, Kerkyacharian & Picard (1996a, 1996c)). Another way of explaining the performance of wavelet shrinkage introduced by D. Donoho and I. Johnstone is the concept of an "oracle". This can be explained as follows. Suppose we want to estimate a quantity μ, with n observations. For that we have a family of estimators $\hat{\mu}_t$ depending on a "tuning" parameter t. A typical example of this situation is to estimate a density f using a kernel method with the tuning parameter being the size of the "window" h. We would be extremely fortunate if every time we have to estimate the quantity μ, comes an oracle telling which t to choose for this precise μ to attain the ideal risk $R(\text{or}, \mu) = \min_t E_\mu ||\hat{\mu}_t - \mu||^2$. We say that we have an oracle inequality for an estimator $\hat{\mu}$ if:

$$E_\mu ||\hat{\mu} - \mu||^2 \leq K_n \left[R(\text{or}, \mu) + \frac{1}{n} \right].$$

This is saying that up to the coefficient K_n, the estimator $\hat{\mu}$ is behaving as if it has an oracle. Consider the Gaussian white noise model, and put $\beta_{jk} = (f, \psi_{jk}), \hat{\beta}_{jk} = \int \psi_{jk}(s) dY(s)$ and consider estimators $\hat{t}$ of the form $\sum_{j \leq j_1} \sum_k \gamma_{jk} \hat{\beta}_{jk} \psi_{jk}$ where γ_{jk} is non stochastic and belongs to $\{0, 1\}$. In this case knowing parameter t consists in knowing where $\gamma_{jk} = 1$, i.e. which coefficient to estimate. It is easily seen that

$$E||\hat{t} - f||_2^2 = \sum_{j \leq j_1} \sum_k \{\gamma_{jk} \frac{1}{n} + (1 - \gamma_{jk}) \beta_{jk}^2\} + \sum_{j > j_1} \sum_k \beta_{jk}^2.$$

Here the oracle has only to tell the places where $j \leq j_1$ and $\beta_{jk}^2 \leq \frac{1}{n}$, to attain $R(\text{or}, f)$. It can be proved, that soft thresholding for example satisfies an oracle inequality with $K_n = (1 + 2 \log n)$. For more discussion of the wavelet oracle see Hall, Kerkyacharian & Picard (1996b).

11.7 Bibliographic remarks

Since the subject "Wavelets and Statistics" is growing rapidly in the moment, it is difficult to provide an up-to-date bibliography that will not be outdated in a short time. Nevertheless, we believe that a brief review of the guidelines in this field will be helpful for the reader. To our knowledge Doukhan (1988) and Doukhan & Leon (1990) were the first to use wavelets in statistics. They introduced the linear wavelet density estimator, and studied its quadratic deviation.

The connection between linear wavelet estimators and Besov spaces appeared in Kerkyacharian & Picard (1992, 1993), Johnstone, Kerkyacharian & Picard (1992). In the same time D. Donoho and I. Johnstone developed the theory of thresholding in a general framework. Their results were published later in Donoho & Johnstone (1994b), Donoho (1994), and Johnstone (1994). Further study in this direction appears in a series of papers by David Donoho and contributors: Donoho (1992a, 1992b, 1993, 1995), Donoho & Johnstone (1994a, 1991, 1995, 1996), Donoho, Johnstone, Kerkyacharian & Picard (1995, 1996, 1997).

Among other contributions which were not discussed in this book, we mention the following works. Antoniadis (1994) and Antoniadis, Grégoire & McKeague (1994) proved the asymptotic normality of the linear wavelet density estimates and investigated different forms of soft thresholding. Fan (1994) and Spokoiny (1996) investigated the use of wavelet thresholding in hypothesis testing. Hall & Patil(1995a, 1995b, 1996b, 1996a) studied the behavior of non linear wavelet estimators in various situations and proved their local adaptivity. These estimators adapt to changing local conditions (such as discontinuity, high oscillations, etc.) to the extent of achieving (up to a log term) the same rate as the optimal linear estimator. Johnstone & Silverman (1997) investigated wavelet regression estimators in the case of stationary correlated noise. Wang (1996) treated the long memory noise setting. Nason (1996), Neumann & Spokoiny (1995) implemented crossvalidation algorithms on thresholding estimates. Marron, Adak, Johnstone, Neumann & Patil (1995) develop the exact risk analysis to understand the small sample behavior of wavelet estimators with soft and hard thresholding. More discussion on wavelet shrinkage mechanism is provided by Bruce & Gao (1996b). For other various aspects of wavelets in statistics see the collection of papers Antoniadis & Oppenheim (1995) and the book of Ogden (1997).

Chapter 12

Computational aspects and statistical software implementations

12.1   Introduction

In this chapter we discuss how to compute the wavelet estimators and give a brief overview of the statistical wavelets software.

There is a variety of software implementations available. One software implementation is **Wavelab.600**, a MATLAB software for wavelet and time frequency analysis. It was written by Buckhut, Chen, Donoho, Johnstone and Scargh and is available on the Internet via

`wavelab @ playfair.stanford.edu` .

There are S-Plus wavelet modules available on `statlib`. They describe how to use the S-Plus Wavelets module, S+ WAVELETS and includes detailed descriptions of the principal S+ WAVELETS functions. It is based on either a UNIX or a Windows system. The intended audience are engineers, scientists and signal analysts, see Oppenheim & Schafer (1975). A recent book on wavelet analysis with S-plus is Bruce & Gao (1996a), see also Nason & Silverman (1994).

A recent interactive user interface in MATLAB is the wavelet TOOLBOX, see Misiti, Misiti, Oppenheim & Poggi (1996). It allows selection of bases and color aided thresholding of one and two dimensional signals.

215

In this chapter we present the software implementation in **XploRe**. The wavelet analysis presented here may be tried using the JAVA interface of XploRe. The macros used for this book are available on the internet via

http: // www.xplore-stat.de.

There is a WWW and a dynamic Java interface available.

Other references on computational aspects are Strang & Nguyen (1996), Young (1993), Foufoula-Georgiou & Kumar (1994) and Burke-Hubbard (1995).

12.2   The cascade algorithm

In this section we present some recursive formulas for wavelet coefficients that allow to compute sequentially the higher level coefficients from the lower level ones and vice versa. These recursions are called *cascade algorithm* (or pyramidal algorithm). They were proposed by Mallat (1989).

First, we define the cascade algorithm for the wavelet coefficients $\alpha_{jk} = (f, \varphi_{jk})$ and $\beta_{jk} = (f, \psi_{jk})$ of a given function f. It will be assumed throughout that we deal only with the bases of compactly supported wavelets constructed starting from a function $m_0(\xi) = \frac{1}{\sqrt{2}} \sum_k h_k e^{-ik\xi}$ (see Chapters 5 - 7), where h_k are real-valued coefficients such that only a finite number of h_k are non-zero. This assumption is satisfied for Daubechies' bases, coiflets and symmlets.

Lemma 5.4 implies that the coefficients α_{jk} and β_{jk} satisfy, for any $j, k \in \mathbb{Z}$, the relations

$$\alpha_{jk} = \sum_l h_{l-2k}\, \alpha_{j+1,l}, \tag{12.1}$$

and

$$\beta_{jk} = \sum_l \lambda_{l-2k}\, \alpha_{j+1,l}, \tag{12.2}$$

where $\lambda_k = (-1)^{k+1} h_{1-k}$ and $\{h_k\}$ are the coefficients of the trigonometric polynomial $m_0(\xi)$. In fact, (5.13) yields

$$
\begin{aligned}
\beta_{jk} &= 2^{j/2} \int f(x)\psi(2^j x - k)dx = \\
&= 2^{(j+1)/2} \sum_s \lambda_s \int f(x)\varphi(2(2^j x - k) - s)dx
\end{aligned}
$$

$$= 2^{(j+1)/2} \sum_s \lambda_s \int f(x)\varphi(2^{j+1}x - 2k - s)dx$$

$$= \sum_s \lambda_s \alpha_{j+1,s+2k} = \sum_l \lambda_{l-2k}\alpha_{j+1,l}.$$

This gives (12.2). The relation (12.1) is obtained similarly, with the use of (5.14).

Together (12.1) and (12.2) define the cascade algorithm. The transformation given by (12.1) is a low-pass filter, while (12.2) is a high-pass filter (see Daubechies (1992), Section 5.6, for explanation of the filtering terminology). Assume that f is compactly supported. Then, as we deal with the bases of compactly supported wavelets, only a finite number of coefficients α_{jl} are non-zero on each level j. Consequently, if the vector of coefficients $\mathbf{y} = \{\alpha_{j_1 l}\}$ for the level j_1 is given, one can reconstruct recursively the coefficients α_{jk}, β_{jk} for levels $j \leq j_1$, by use of linear recursive formulas (12.1), (12.2). Note that, under our assumption on the finiteness of the vector h_k, the number of non-zero coefficients α_{jk}, β_{jk} decreases with the level j, since the discrete convolutions in (12.1) and (12.2) are sampled at points $2k$. If the procedure (12.1), (12.2) stops at level j_0, the resulting vector of wavelet coefficients $\mathbf{w} = (\{\alpha_{j_0 k}\}, \{\beta_{j_0 k}\}, \dots, \{\beta_{j_1-1,k}\})^T$ can be presented as

$$\mathbf{w} = \mathcal{W}\mathbf{y}, \tag{12.3}$$

where $\mathcal{W}$ is a matrix.

It is possible to invert the cascade algorithm and thus to get the values of coefficients $\mathbf{y}$, starting from $\mathbf{w}$. The inverse algorithm can be presented by the following recursive scheme:

$$\alpha_{j+1,s} = \sum_k h_{s-2k}\alpha_{jk} + \sum_k \lambda_{s-2k}\beta_{jk}, \tag{12.4}$$

running from $j = j_0$ to $j = j_1 - 1$. To get (12.4) directly, observe that $\alpha_{j+1,s} = (P_{V_{j+1}}(f), \varphi_{j+1,s})$, where $P_{V_{j+1}}(f)$ is the orthogonal projection of f on the space V_{j+1}. Therefore, applying (3.6), we get

$$\alpha_{j+1,s} = \sum_k \alpha_{jk}(\varphi_{jk}, \varphi_{j+1,s})$$
$$+ \sum_k \beta_{jk}(\psi_{jk}, \varphi_{j+1,s}). \tag{12.5}$$

But, in view of (5.14),

$$(\varphi_{jk}, \varphi_{j+1,s}) = \sum_l h_l \int \varphi_{j+1,2k+l}\, \varphi_{j+1,s} = \sum_l h_l \delta_{2k+l,s} = h_{s-2k},$$

and, similarly,

$$(\psi_{jk}, \varphi_{j+1,s}) = \lambda_{s-2k}.$$

These relations and (12.5) yield (12.4).

Now we turn to the empirical wavelet coefficients $\hat\alpha_{jk}, \hat\beta_{jk}$. The cascade algorithm applies to them as well. However, there are some modifications that we are going to discuss.

First, observe that in the statistical estimation setup (see Chapter 10) the aim is to compute not only the empirical wavelet coefficients, but the wavelet estimator at gridpoints $z_1, \ldots, z_m$, i.e. the vector

$$\mathbf{f} = (f_1, \ldots, f_m),$$

with

$$f_l = \sum_k \hat\alpha_{j_0 k}\varphi_{j_0 k}(z_l) + \sum_{j=j_0}^{j_1} \sum_k \eta_{jk}(\hat\beta_{jk})\psi_{jk}(z_l), \ \ l = 1, \ldots, m, \qquad (12.6)$$

where

$$\hat\alpha_{jk} = \frac{1}{m}\sum_{i=1}^{m} \hat y_i \varphi_{jk}(z_i), \qquad (12.7)$$

$$\hat\beta_{jk} = \frac{1}{m}\sum_{i=1}^{m} \hat y_i \psi_{jk}(z_i) \qquad (12.8)$$

(cf. (10.10) - (10.12)). Here $\hat y_i$ are the binned data and $\eta_{jk}(\cdot)$ are some known functions (thresholding transformations, cf. Section 11.2). We assume that z_i are mapped in $[0, 1]$, so that $z_i = \frac{i}{m}$. The difference between density and nonparametric regression settings appears only in the definition of the binned values $\hat y_i, i = 1, \ldots, m$. For the density case $\hat y_j$ are the values of a histogram, while for the nonparametric regression case they are the values of a regressogram (see Section 10.8). The estimator (12.6) - (12.8) can be used for other nonparametric settings as well, with a proper definition of the binned values $\hat y_i$.

Computation of the estimator (12.6) - (12.8) is not an easy task: in fact, usually the functions φ_{jk}, ψ_{jk} are not available in an explicit form (see Chapters 5-7). We will see below that the cascade algorithm allows a recursive computation of the empirical wavelet coefficients $\hat{\alpha}_{jk}, \hat{\beta}_{jk}$, $j_0 \le j \le j_1$. The question about the efficient computation of the values of the estimator $f_1, \ldots, f_m$ is more delicate. We defer it to the next section where we present some fast (but approximate) methods for such computation commonly used in practice.

To get the empirical cascade algorithm observe that the empirical wavelet coefficients can be written as

$$\hat{\alpha}_{ik} = (q_m, \varphi_{jk}),$$
$$\hat{\beta}_{ik} = (q_m, \psi_{jk}),$$

where q_m is the measure

$$q_m = \frac{1}{m} \sum_{i=1}^{m} \hat{y}_i \delta_{\{z_i\}},$$

with $\delta_{\{x\}}$ being the Dirac mass at point x, and $(q_m, \varphi_{jk}) = \int \varphi_{jk} dq_m$. Analogously to (12.1) and (12.2) (but replacing $f(x)dx$ by dq_m in the calculations) we get the following recursive formulae

$$\hat{\alpha}_{jk} = \sum_{l} h_{l-2k} \hat{\alpha}_{j+1,l} = \sum_{l} h_l \hat{\alpha}_{j+1,l+2k}, \qquad (12.9)$$

$$\hat{\beta}_{jk} = \sum_{l} \lambda_{l-2k} \hat{\alpha}_{j+1,l} = \sum_{l} \lambda_l \hat{\alpha}_{j+1,l+2k}, \qquad (12.10)$$

Thus, to compute $\hat{\beta}_{jk}, \hat{\alpha}_{jk}$, for $j_0 \le j \le j_1$, we start with the computation of $\hat{\alpha}_{j_1 k} = \frac{1}{m} \sum_{i=1}^{m} \hat{y}_i \varphi_{j_1 k}(z_i)$, (i.e. start with the highest level $j = j_1$), and then obtain the values $\hat{\beta}_{jk}, \hat{\alpha}_{jk}$ recursively from (12.9) - (12.10), level by level, up to $j = j_0$. Clearly, (12.9) - (12.10) is the "empirical" version of the cascade algorithm (12.1) - (12.2). The coefficients $\{h_k\}$ are tabulated in Daubechies (1992), for common examples of compactly supported father and mother wavelets (see also Appendix A). Note that for such common wavelets the number of non-zero coefficients $\{h_k\}$ or $\{\lambda_k\}$ does not exceed 10-20.

A problem with the implementation of (12.9) - (12.10) is that the initial values $\hat{\alpha}_{j_1 k}$ are not easy to compute, again for the reason that the functions $\varphi_{j_1 k}$ are not explicitly known.

The formulas (12.9) - (12.10) that define the empirical cascade algorithm are the same as those for the original cascade algorithm (12.4) - (12.5); the only difference is in the definition of the starting values: $\{\alpha_{j_1 k}\}$ are replaced in (12.9) - (12.10) by $\{\hat{\alpha}_{j_1 k}\}$. By analogy to the previous argument, it could seem that the inverse algorithm should be also given by the recursion (12.7):

$$\hat{\alpha}_{j+1,s} = \sum_k h_{s-2k}\,\hat{\alpha}_{jk} + \sum_k \lambda_{s-2k}\hat{\beta}_{jk}. \tag{12.11}$$

However, this is not exactly the case, because we operate with the empirical measure q_m, and not with a function $f \in L_2(\mathbb{R})$. The fact that α_{jk}, β_{jk} are wavelet coefficients of such a function f was essential to show (12.7). The empirical cascade algorithms (12.9) - (12.10) and (12.11) act on finite discrete arrays of coefficients, and, in general, (12.11) is not the exact inversion of (12.9) - (12.10). To get the exact inversion it suffices to modify (12.9) - (12.10) and (12.11) by introducing periodic extensions of the computed coefficients onto $\mathbb{Z}$, along with dyadic summations. This constitutes the technique of *discrete wavelet transform (DWT)*, see Mallat (1989). We describe it in the next section. Note beforehand that the use of inverse algorithm is fundamental for the computation. In fact, the idea is to run the forward algorithm until $j = j_0$, then to apply a thresholding transformation to the obtained wavelet coefficients, and to run the inverse algorithm, starting from these transformed coefficients, until $j = K$. The output of this procedure is claimed to give approximately the values $f_1, \ldots, f_m$ of the wavelet estimator at the gridpoints.

12.3   Discrete wavelet transform

To define the *DWT* we first introduce some linear transformations. now. For $l \in \mathbb{Z}, r \in \mathbb{Z}$, and an integer s denote $(l + r)$ mod s the mod s sum of l and r. Let $\mathbf{Z} = (Z(0), \ldots, Z(s-1))$ be a vector where s is an even integer. Define the transformations $\mathcal{L}^s$ and $\mathcal{H}^s$ of the vector $\mathbf{Z}$ coordinatewise, for $k = 0, \ldots, s/2 - 1$, by

$$\mathcal{L}^s Z(k) \;=\; \sum_l h_l Z((l + 2k) \bmod s),$$

$$\mathcal{H}^s Z(k) = \sum_l \lambda_l Z((l + 2k) \bmod s).$$

These are the analogues of the low-pass filter (12.1) and the high-pass filter (12.2) respectively, with the mod s addition that can be also interpreted as a periodic extension of data. Clearly, $\mathcal{L}^s$ and $\mathcal{H}^s$ map the vector $\mathbf{Z}$ of dimension s on two vectors $\mathcal{L}^s\mathbf{Z}$ and $\mathcal{H}^s\mathbf{Z}$ of dimension $s/2$.

The DWT acts by iterative application of the transformations $\mathcal{L}$ and $\mathcal{H}$. It starts from the initial vector $\left((Z(0), \ldots, Z(2^K - 1)\right)$ which we denote for convenience in the following way as the two entries array:

$$\{\alpha(K, k), k = 0, \ldots, 2^K - 1\}.$$

The DWT computes recursively the vectors

$$\{\alpha(j, k), k = 0, \ldots, 2^j - 1\}, \quad \{\beta(j, k), k = 0, \ldots, 2^j - 1\}$$

for $0 \le j \le K - 1$. The recursions defining the DWT are:

$$\alpha(j, k) = \mathcal{L}^{2^{j+1}}\alpha(j + 1, k) = \sum_l h_l \alpha(j + 1, (l + 2k) \bmod 2^{j+1}), \qquad (12.12)$$

$$\beta(j, k) = \mathcal{H}^{2^{j+1}}\alpha(j + 1, k) = \sum_l \lambda_l \alpha(j + 1, (l + 2k) \bmod 2^{j+1}). \qquad (12.13)$$

Remark that the notation $\alpha(j, k)$, $\beta(j, k)$ is reminiscent of the wavelet coefficients $\alpha_{j,k}$, $\beta j, k$, while the above recursions are similar to the cascade algorithm. However, we would like to emphasize that the definition of the DWT is given irrespectively of the framework of the previous section: in fact, the DWT is just a composition of linear orthogonal transformations presented by the recursions (12.12) and (12.13). The reason for adopting such a notation is that in the next section, where we consider statistical applications of the DWT, the values $\alpha(j, k)$, $\beta(j, k)$ will approximately correspond to $\alpha_{j,k}$, $\beta j, k$.

Observe that the recursions (12.12) and (12.13) can be used to define $\alpha(j, k)$ and $\beta(j, k)$ not only for $k = 0, \ldots, 2^j - 1$, but also for all $k \in \mathbb{Z}$. It follows from (12.12) and (12.13) that such extended sequences are periodic:

$$\alpha(j, k) = \alpha(j, k + 2^j), \quad \beta(j, k) = \beta(j, k + 2^j), \quad \forall\, k \in \mathbb{Z}.$$

The inverse DWT is defined similarly to (12.11), but with the periodically extended data. It starts from the vectors

$$\{\alpha(j_0, k), k = 0, \ldots, 2^{j_0} - 1\}, \quad \{\beta(j_0, k), k = 0, \ldots, 2^{j_0} - 1\}$$

whose periodic extensions are denoted

$$\{\tilde{\alpha}(j_0, k), k \in \mathbb{Z}\}, \quad \{\tilde{\beta}(j_0, k), k \in \mathbb{Z}\}$$

and computes in turn the vectors $\{\alpha(j, s), s = 0, \ldots, 2^j - 1\}$, until the level $j = K - 1$, following the recursions:

$$\tilde{\alpha}(j+1, s) = \sum_k h_{s-2k}\, \tilde{\alpha}(j, k) + \sum_k \lambda_{s-2k}\tilde{\beta}(j, k), \quad s \in \mathbb{Z}, \qquad (12.14)$$

$$\alpha(j+1, s) = \tilde{\alpha}(j+1, s), \quad s = 0, \ldots, 2^{j+1} - 1. \qquad (12.15)$$

Clearly, (12.14) implies the periodicity of all intermediate sequences:

$$\tilde{\alpha}(j+1, s) = \tilde{\alpha}(j+1, s + 2^{j+1}), \quad s \in \mathbb{Z}.$$

12.4   Statistical implementation of the DWT

Binning

The computation of wavelet estimators is based on the DWT described above. The DWT needs to work on signals of length $m = 2^K$, where K is an integer. In applications the sample size is often not a power of 2. The data needs therefore to be transformed to a grid of $m = 2^K$ equispaced points. This is true both for density estimation and regression smoothing. The binning procedures for the density and regression wavelet estimation were introduced in Sections 10.2 and 10.8 respectively. Here we would like to discuss the effect of binning with different bin size on the quality of wavelet estimators.

We investigate again the example of density estimation already considered in Chapter 10, Figures 10.1–10.11. For our example of $n = 500$ data points we have investigated the binning into $m = 8, 16, 32, 64, 256, 512$ binpoints. The corresponding estimated ISE values are given in Table 12.1.

bins	S8 hard	S8 soft	H hard	H soft
8	1	1.4157	1	1.4335
16	0.29267	1.0596	0.13811	0.55132
32	0.054237	0.26103	0.047822	0.41557
64	0.053587	0.23887	0.029666	0.22516
128	0.068648	0.27802	0.057907	0.29147
256	0.15012	0.37995	0.1348	0.37757
512	0.19506	0.53409	0.18746	0.55368

Table 12.1: ISE values for different bin sizes

One sees that the ISE values have a minimum at $m = 64 = 2^K, K = 6$. The corresponding ISE curve for $S8$ are given in Figure 12.1. Although there is an "optimal" bin size we must be careful in interpreting it in a statistical way. The binning is merely a presmoothing and was not taken into account in the theoretical calculations e.g. in Chapter 10. The higher the number of bins the more we loose the computational efficiency. The values in Figure 12.1 represent thus more a trade off between computational speed and presmoothing.

Approximate computation of wavelet estimators

The implementation of DWT for an approximate computing of statistical estimators (12.6) - (12.8) follows the next scheme.

(i) *Limits of the computation and initial values.* Instead of starting at the level j_1, the algorithm (12.12) - (12.13) starts at $j = K = \log_2 m$. The initial values $\alpha(K, l)$ are set to be equal to the binned observations:

$$\alpha(K, l) := \hat{y}_{l+1}, \quad l = 0, \ldots, m - 1.$$

(ii) *Forward transform.* The DWT (12.12) - (12.13) runs from $j = K$ until $j = j_0$, and results in the vector of coefficients

$$\hat{\mathbf{w}} = (\{\alpha(j_0, k)\}, \{\beta(j_0, k)\}, \{\beta(j_0 + 1, k)\}, \ldots, \{\beta(K - 1, k)\})^T.$$

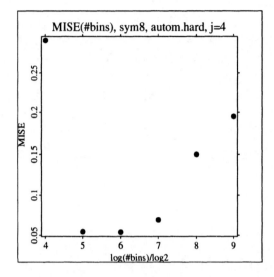

Figure 12.1: ISE for $S8$ as a function of bin size

The vectors $\{\alpha(j,k)\}$, $\{\beta(j,k)\}$ are of length 2^j, and thus $\hat{\mathbf{w}}$ is of length 2^K.

(iii) *Inverse transform.* The inverse DWT (12.14) - (12.15) runs from $j = j_0$ until $j = K - 1$, starting with the vector of thresholded initial values

$$\mathbf{w}^* = (\{\alpha^*(j_0,k)\}, \{\beta^*(j_0,k)\}, \{\beta^*(j_0+1,k)\}, \ldots, \{\beta^*(K-1,k)\})^T$$

where

$$\alpha^*(j_0,k) = \alpha(j_0,k), \quad \beta^*(j,k) = \eta_{jk}(\beta(j,k)). \qquad (12.16)$$

The inverse DWT results in $2^K = m$ values $\{\alpha^*(K,l), l = 0, \ldots, m-1\}$. ((The output is the vector $\mathbf{f}^* = (f_1^*, \ldots, f_m^*)^T$, where

$$f_{l+1}^* := \alpha^*(K,l), \, l = 0, \ldots, m-1.$$

The values f_l^* are taken as approximations for f_l.

Some remarks about this algorithm are immediate. First, the very definition of the DWT comprises a periodic extension of the data at any step of the method. This is a consequence of the dyadic summation. For example, on the first step the original values $\hat{y}_k$ are regarded as being periodically extended on $\mathbb{Z}$, with period $m = 2^K$, so that $\hat{y}_{k+m} = \hat{y}_k$, $k \in \mathbb{Z}$.

Next, we comment on the fact that the upper level j_1 does not appear in the description of the algorithm (i) – (iii). In pactice one usually sets $j_1 = K$, and applies the hard or soft thresholding to all the coefficients on the levels $j = j_1, \ldots, K - 1$ (the level K is not thresholded since it contains only the α coefficients). However, if one wants to exclude the coefficients of the levels $> j_1$, as for example in the linear wavelet estimator, the definition (12.16) yields this possibility by setting

$$\eta_{jk}(u) \equiv 0, \quad j_1 < j \le K.$$

Similarly to (12.3), one can present the algorithm (i) – (iii) in the matrix form. Let $\hat{\mathbf{y}} = (\hat{y}_1, \ldots, \hat{y}_m)^T$. Then the result of the forward transform is

$$\hat{\mathbf{w}} = \hat{\mathcal{W}}\hat{\mathbf{y}}, \tag{12.17}$$

where $\hat{\mathcal{W}}$ is a $m \times m$ matrix. One can show that $\hat{\mathcal{W}}$ is an orthogonal matrix, since it can be presented as a product of finite number of orthogonal matrices corresponding to the steps of the algorithm (Mallat (1989)). Denote $\mathcal{T}$ the thresholding transformation (12.16):

$$\mathbf{w}^* = \mathcal{T}(\hat{\mathbf{w}}).$$

The inverse DWT is defined by the inverse matrix $\hat{\mathcal{W}}^{-1}$ and, in view of the orthogonality, $\hat{\mathcal{W}}^{-1} = \hat{\mathcal{W}}^T$. Hence, the output $\mathbf{f}^* = (f_1^*, \ldots, f_m^*)^T$ of the method (i) – (iii) is

$$\mathbf{f}^* = \hat{\mathcal{W}}^T \mathbf{w}^* = \hat{\mathcal{W}}^T \mathcal{T}(\hat{\mathcal{W}}\hat{\mathbf{y}}).$$

If we deal with linear wavelet estimators and j_1 takes the maximal value: $j_1 = K - 1$, then $\mathcal{T}$ is the identity transformation and we get $\mathbf{f}^* = \hat{\mathbf{y}}$. This is natural: if all the coefficients for all levels are present the estimator reproduces the data.

The method (i) – (iii) is commonly used for computation of wavelet estimators. It is faster than the fast Fourier transform: it requires only $O(m)$ operations. However, except for the case of linear Haar wavelet estimator, *it does not compute* the estimator (12.6), but rather an approximation. This fact is not usually discussed in the literature.

Let us give an intuitive argument explaining why the output $f_1^*, \ldots, f_m^*$ of the method (i) - (iii) approximates the values $f_1, \ldots, f_m$ of the estimator (12.6). Consider only the linear wavelet estimator (i.e. put $\eta_{jk}(u) = u$, $\forall j_0 \le j \le j_1, k$). Assume for a moment that the initial values $\hat{y}_l$ of the method (i) - (iii) satisfy

$$\hat{y}_l \approx \sqrt{m}\hat{\alpha}_{Kl} = 2^{K/2}\hat{\alpha}_{Kl}. \tag{12.18}$$

We know that the recursions (12.9) - (12.10) compute the values $\hat{\alpha}_{jk}, \hat{\beta}_{jk}$, and that the forward transform (12.12) - (12.13) does approximately the same job, if the initial values are the same. If (12.18) holds, the initial values of (12.12) - (12.13) in (iii) differ from those of the recursions (12.9) - (12.10) approximately by the factor $\sqrt{m}$. The linearity of recursions entails that the outputs of these forward transforms differ by the same factor, i.e.

$$\alpha(j_0, k) \approx \sqrt{m}\hat{\alpha}_{j_0 k}, \quad \beta(j, k) \approx \sqrt{m}\hat{\beta}_{jk}.$$

This and (12.7) - (12.8) yield

$$\hat{\mathbf{w}} \approx \frac{1}{\sqrt{m}}\mathcal{W}_m\hat{\mathbf{y}}, \tag{12.19}$$

where $\mathcal{W}_m$ is the $m \times m$ matrix with columns

$$(\{\varphi_{j_0 k}(z_i)\}, \{\psi_{j_0 k}(z_i)\}, \{\psi_{j_0+1,k}(z_i)\}, \ldots, \{\psi_{K-1,k}(z_i)\})^T, \quad i = 1, \ldots, m.$$

Combining (12.17) and (12.19), we obtain:

$$\hat{W} \approx \frac{1}{\sqrt{m}}\mathcal{W}_m.$$

Now, for linear wavelet estimates $\eta_{jk}(u) = u$ for $j_0 \le j \le j_1$, $\eta_{jk}(u) = 0$ for $j > j_1$, and thus the thesholding transformation $\mathcal{T}$ is defined by the idempotent matrix $A = (a_{ij})_{i,j=1,\ldots,m}$, with $a_{ii} = 1$ if $1 \le i \le 2^{j_1+1}$, and $a_{ij} = 0$ otherwise. Therefore,

$$\mathbf{f}^* = \hat{W}^T A \hat{W}\hat{\mathbf{y}} \approx \frac{1}{m}\mathcal{W}_m^T A \mathcal{W}_m\hat{\mathbf{y}} = \mathbf{f}, \tag{12.20}$$

where the last equality is just the vector form of (12.6). This is the desired approximation.

It remains to explain why (12.18) makes sense. We have

$$\sqrt{m}\hat{\alpha}_{Kl} = \frac{1}{\sqrt{m}} \sum_{i=1}^{m} \hat{y}_i \varphi_{Kl}(z_i) = \sum_{i=1}^{m} \hat{y}_i \varphi(i-l).$$

Hence for the Haar wavelet (12.18) holds with the exact equality: $\sqrt{m}\hat{\alpha}_{Kl} = \hat{y}_l$. For coiflets we have $\alpha_{jl} \approx 2^{-j/2} f\left(\frac{l}{2^j}\right)$ with a precision $O(2^{-j\nu})$ where ν is large enough, since a number of first moments of father coiflet φ vanish (note that this is true only if f is smooth enough). With some degree of approximation, one could extend this to the empirical values: $\hat{\alpha}_{Kl} \approx 2^{-K/2}\hat{y}_l$ which gives (12.18). For general wavelet bases (12.18) is not guaranteed and the above intuitive argument fails. Donoho (1992b) and Delyon & Juditsky (1996b) discuss this issue in more detail and characterize specific wavelet bases that guarantee the relation $\alpha_{jl} \approx 2^{-j/2} f\left(\frac{l}{2^j}\right)$ with a precision $O(2^{-j\nu})$ where ν is large enough.

REMARK 12.1 In general, one cannot claim that the approximation of the estimator (12.6) - (12.8) given by the DWT based algorithm (i) - (iii) is precise. The above intuitive argument is fragile in several points.

First, it relies on (12.18) which is difficult to check, except for some special cases, such as the Haar wavelet basis.

Second, it assumes the equivalence of (12.12) - (12.13) and (12.9) - (12.10) which is not exactly the case in view of the dyadic summation (which means also the periodic extension, as mentioned above). The periodic extension is perfect if the estimated function f itself can be extended periodically on $I\!R$ without loss of continuity. Otherwise the quality of estimation near the endpoints of the interval becomes worse. Several suggestions are possible to correct this: the most useful is mirroring (see Section 10.8). With mirroring the new vector of data $\hat{y}$ has the dimension $2m$ and the new values $\hat{y}_l$ are not independent even for the i.i.d. regression or Gaussian white noise models.

Third, the intuitive argument leading to (12.20) was presented only for linear wavelet estimators. With a nonlinear transformation $\mathcal{T}$ it should be modified and becomes even more fragile. But it is likely that with hard or soft thresholding the argument holds through: these transformations are linear on the entire set where they do not vanish.

Finally, as mentioned above the approximation makes sense only if f is smooth enough.

With these remarks and the fact that the *DWT* based estimators are almost the only computational tool that works well in practice, we conclude that it is important to study the statistical properties of these estimators directly. Donoho & Johnstone (1995) undertake such a study for the Gaussian white model. We are not aware of similar studies for other models. In general, the nice statistical results obtained for estimators (12.6) - (12.8) are not sufficient to justify the practical procedures. Moreover, even for the estimators (12.6) - (12.8) the results are not always complete, because they do not account for the effect of binning. These problems remain open.

REMARK 12.2 In general, the bases of compactly supported wavelets are defined with $h_k \neq 0$ for $k \in [N_0, N_1]$, see Chapters 6 and 7. However, in simulations one often shifts h_k to get $N_0 = 0$; thus the support of $\{h_k\}$ becomes the set of integers $k \in [0, N_1 - N_0]$. Note that the resulting wavelet estimator is different from the original one. For Daubechies' wavelets $N_0 = 0$ and this discussion does not arise.

If one uses the linear wavelet estimator, the conditions of vanishing moments are preserved under the shift of coefficients $\{h_k\}$. A significant difference appears only near boundaries or jumps. For nonlinear thresholded case it is clear that the wavelet estimators for shfted and non-shifted situations are different.

12.5   Translation invariant wavelet estimation

In spite of a nice mathematical theory, simulations show that in the neighborhood of discontinuities the wavelet estimators can exhibit pseudo-Gibbs phenomena. Of course, these phenomena are much less pronounced than in the case of Fourier series estimators where they are of global nature and of larger amplitude. However, they are present in wavelet estimators. Here we are going to explain how to reduce these effects.

The idea of improvement is based on the fact that the size of pseudo-Gibbs phenomena depends mainly on the location of a discontinuity in the data. For example, when using the Haar wavelets, a discontinuity located at $m/2$ gives no Gibbs oscillations; a discontinuity near $m/3$ leads to significant pseudo-Gibbs effects. Roughly speaking, the amplitude of pseudo-Gibbs oscillations is proportional to the square root of the number of wavelet coefficients affected by the discontinuity (if a wavelet coefficient is affected by a discontinuity, the thresholding procedure does not suppress noise in the empirical wavelet coefficient). In case of a discontinuity at $m/3$ approximately $\log m$ wavelet coefficients are affected by the discontinuity.

A possible way to correct this misalignment between the data and the basis is to shift the data so that their discontinuities change the position. Hopefully, the shifted signal would not exhibit the pseudo-Gibbs phenomena. After thresholding the estimator can be unshifted.

Unfortunately, we do not know the location of the discontinuity. One reasonable approach in this situation is optimization: introduce some qualitative measure of artifacts and minimize it by a proper choice of the shift. But if the signal has several discontinuities they may interfere with each other. That means that the best shift for one discontinuity may be the worst for another discontinuity. This undermines the idea of optimization with respect to shifts in general situations.

Another, more robust, approach is based on the technique called *stationary wavelet transform*. From an engineering point of view this transform is discussed by Rioul & Vetterli (1991) and Pesquet, Krim & Carfantan (1994). Statistical applications of stationary wavelet transform are presented in Coifman & Donoho (1995) and used also by Nason & Silverman (1994). The corresponding statistical estimator is called translation invariant wavelet estimator.

The basic idea is very simple. As above, consider the problem of estimating the vector of values $(f(z_1), \ldots, f(z_m))$ of an unknown function f (probability density, regression, etc.) at the gridpoints $z_1, \ldots, z_m$.

Suppose that we are given the binned data $\hat{\mathbf{y}} = (\hat{y}_1, \ldots, \hat{y}_m)^T, m = 2^K$. Define the shift operator

$$S_\tau^m \hat{\mathbf{y}} = (\hat{y}_{\tau+1}, \ldots, \hat{y}_{\tau+m})^T,$$

where τ is an integer and, by periodic extension, $\hat{y}_{i-m} = \hat{y}_{i+m} = \hat{y}_i, i = 1, \ldots, m$. The *translation invariant wavelet estimator* is the vector $\mathbf{f}^{TI} = (f_1^{TI}, \ldots, f_m^{TI})$ defined as follows:

$$\mathbf{f}^{TI} = \frac{1}{m} \sum_{\tau=0}^{m-1} S_{-\tau}^m \hat{\mathcal{W}}^T \mathcal{T}(\hat{\mathcal{W}} S_\tau^m \hat{\mathbf{y}}), \qquad (12.21)$$

where $\hat{\mathcal{W}}$ is the matrix of the discrete wavelet transform (DWT).

In words, we do the following:

(i) for any feasible shift we calculate the DWT of the shifted data, threshold the result, invert the DWT and unshift the signal;

(ii) finally we average over all the shifts.

Since the computation of each summand in (12.21) takes $O(m)$ operations, at first glance it seems that $\mathbf{f}^{TI}$ needs the $O(m^2)$ operations. Fortunately, there exists an algorithm requiring only $O(m \log m)$ operations. Let us explain how it works.

The idea is close to that of the DWT but it involves an additional complication due to the shifts. Introduce the vectors

$$
\begin{array}{ll}
\mathbf{v}_1 = \{\alpha(K-1, k)\}_{k=0}^{m/2-1} & , \quad \mathbf{w}_1 = \{\beta(K-1, k)\}_{k=0}^{m/2-1}, \\
\mathbf{v}_2 = \{\alpha(K-2, k)\}_{k=0}^{m/4-1} & , \quad \mathbf{w}_2 = \{\beta(K-2, k)\}_{k=0}^{m/4-1},
\end{array}
$$

$$
\vdots \qquad\qquad\qquad\qquad \vdots
$$

$$
\mathbf{v}_K = \alpha(0,0) \qquad\qquad , \quad \mathbf{w}_K = \beta(0,0),
$$

and set $\mathbf{v}_0 = \hat{\mathbf{y}}$. With this notation the first step of the DWT in the method (i) - (iii) of the previous section is

$$
\begin{array}{rcl}
\mathbf{v}_1 & = & \mathcal{L}^m \mathbf{v}_0, \\
\mathbf{w}_1 & = & \mathcal{H}^m \mathbf{v}_0.
\end{array}
$$

The second step is

$$
\begin{array}{rcl}
\mathbf{v}_2 & = & \mathcal{L}^{m/2} \mathbf{v}_1, \\
\mathbf{w}_2 & = & \mathcal{H}^{m/2} \mathbf{v}_1,
\end{array}
$$

etc.

A similar algorithm is used for the fast calculation of $\hat{\mathcal{W}} S_\tau^m \hat{\mathbf{y}}$. The algorithm returns a $m \times \log_2 m$ matrix which we call the TI Table according to Coifman & Donoho (1995). This matrix has the following properties:

(i) for any integer τ, $0 < \tau < n$ it contains $\hat{\mathcal{W}} S_\tau^m \hat{\mathbf{y}}$;

(ii) the TI Table can be computed in $O(\log_2 m)$ operations;

(iii) the extraction of $\hat{W}S_\tau^m\hat{y}$ for a certain τ from the TI Table requires $O(m)$ operations.

We start with

$$\mathbf{v}_{10} = \mathcal{L}^m\mathbf{v}_0, \quad \mathbf{v}_{11} = \mathcal{L}^m S_1^m \mathbf{v}_0,$$
$$\mathbf{w}_{10} = \mathcal{H}^m\mathbf{v}_0, \quad \mathbf{w}_{11} = \mathcal{H}^m S_1^m \mathbf{v}_0.$$

The output data of this first step are $(\mathbf{w}_{10}, \mathbf{w}_{11})$. They constitute the last row in the TI Table. Note that both $\mathbf{w}_{10}$ and $\mathbf{w}_{11}$ are of dimension $m/2$.

At the next step we filter the vector $(\mathbf{v}_{10}, \mathbf{v}_{11})$:

$$\mathbf{v}_{20} = \mathcal{L}^{m/2}\mathbf{v}_{10}, \quad \mathbf{v}_{21} = \mathcal{L}^{m/2} S_1^{m/2}\mathbf{v}_{10},$$
$$\mathbf{v}_{22} = \mathcal{L}^{m/2}\mathbf{v}_{11}, \quad \mathbf{v}_{23} = \mathcal{L}^{m/2} S_1^{m/2}\mathbf{v}_{11},$$

and

$$\mathbf{w}_{20} = \mathcal{H}^{m/2}\mathbf{v}_{10}, \quad \mathbf{w}_{21} = \mathcal{H}^{m/2} S_1^{m/2}\mathbf{v}_{10},$$
$$\mathbf{w}_{22} = \mathcal{H}^{m/2}\mathbf{v}_{11}, \quad \mathbf{w}_{23} = \mathcal{H}^{m/2} S_1^{m/2}\mathbf{v}_{11}.$$

The vectors $(\mathbf{w}_{20}, \mathbf{w}_{21}, \mathbf{w}_{22}, \mathbf{w}_{23})$ give the next row in the TI Table. These are four vectors, each of dimension $m/4$.

After $\log_2 m = K$ iterations we completely fill the TI Table. Then the thresholding transformation $\mathcal{T}$ is applied. Finally, one can invert the TI Table, so that the result of inversion gives the estimator (12.21). The fast inversion algorithm is similar to (12.14) - (12.15). We refer to Coifman & Donoho (1995) for further details.

The translation invariant wavelet density estimation has been shown already in Figure 10.12 for a soft thresholding transformation $\mathcal{T}$. In Figure 12.2 we show the same density example as in Section 10.4 with a hard threshold of $t = 0.25\max|\hat{\beta}_{jk}|$.

12.6 Main wavelet commands in XploRe

The above computational algorithms are implemented in the interactive statistical computing environment XploRe. The software is described in the book Härdle et al. (1995) and is available via the **http://www.xplore-stat.de** address. Here we discuss only the main wavelet commands. In the appendix we give more information about how to obtain the software.

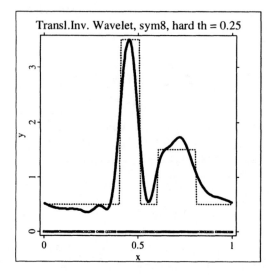

Figure 12.2: Translation invariant density estimation with S8 and hard threshold $0.25 \max |\hat{\beta}_{jk}|$

Wavelet generating coefficients

The XploRe wavelet library implements 22 common basis wavelets. These are the Haar $(= D2)$ wavelet, and

$$D4, \ldots, D8, \ldots, D20; S4, S5, \ldots, S10; C1, C2, \ldots, C5 \qquad (12.22)$$

with the $\{h_k\}$ coefficients from Daubechies (1992). These coefficients are stored in a file

"/data/wavelet.dat"

The letter "D" stands for Daubechies, "S" for symmlet, "C" for coiflet. There are 296 coefficients all together. We list them in Table A.1 in the appendix. This table shows the coefficients in the order given in (12.22). The indices of each coefficient sequence are given in Table A.2. The wavelet Symmlet7 $S7$ is the 14th wavelet and thus the coefficients 139 to 152 are taken out of Table A.1. We list in Table 12.2 the coefficients for $S7$.

The XploRe command is library("wavelet"). This call to the wavelet module automatically yields the $\{h_k\}$ coefficient vectors haar, daubechies4,

139	0.00268181	146	0.536102
140	-0.00104738	147	0.767764
141	-0.0126363	148	0.28863
142	0.0305155	149	-0.140047
143	0.0678927	150	-0.107808
144	-0.0495528	151	0.00401024
145	0.0174413	152	0.0102682

Table 12.2: The coefficients for $S7$.

daubechies8 etc. These are generically denoted by h in the sequel (e.g. h = daubechies4).

The coefficients h are used to generate the discrete wavelet transform via fwt or dwt to generate the functions φ and ψ.

Discrete wavelet transform

Let $K \geq 1$ be the level where the DWT starts, and x be the input vector of length $m = 2^K$ (it corresponds to the vector $\hat{y}$ in the notation of previous sections). Let $0 \leq j < K$ be the level where the DWT stops, and the variable $1 = 2^j$ be the number of father wavelets on this output level j.

The DWT is realized by the following command:

$$\{a, b\} = \texttt{fwt}(x, 1, h),$$

where $\{a, b\}$ is the output vector of dimension 2^K (it corresponds to the vector $\hat{w}$ in the notation of previous sections). It is divided into two subvectors: a, the vector of coefficients $\{\alpha(j, k)\}$, and b, the vector of coefficients

$$(\{\beta(j, k)\}, \{\beta(j+1, k)\}, \ldots, \{\beta(K-1, k)\}).$$

The abbreviation fwt stands for "fast wavelet transform". Alternatively one may use the command

$$y = \texttt{dwt}(x, 1, h)$$

Here y denotes the vector $\hat{w}$.

Consider a numerical example. The command $\mathbf{x} = \#(0,0,1,1)$ would generate a step function. Here $m = 4$, $K = 2$. The command

$$\{\mathbf{a}, \mathbf{b}\} = \mathtt{fwt}(\mathbf{x}, 1, \mathtt{haar})$$

would result in this case in

$$\begin{aligned} \mathbf{a} &= \alpha(0,0) = 1/2, \\ \mathbf{b} &= (\beta(0,0), \beta(1,0), \beta(1,1)) = (1/2, 0, 0). \end{aligned}$$

(Here the output level is $j = 0$.) It is easy to check this result directly, starting from the values $\alpha(2,0) = \alpha(2,1) = 0, \alpha(2,2) = \alpha(2,3) = 1$ and using the particular form that takes the DWT (12.12) - (12.13) for the Haar wavelet:

$$\begin{aligned} \alpha(1,0) &= h_0\alpha(2,0) + h_1\alpha(2,1) = \frac{1}{\sqrt{2}}(\alpha(2,0) + \alpha(2,1)), \\ \alpha(1,1) &= h_0\alpha(2,2) + h_1\alpha(2,3) = \frac{1}{\sqrt{2}}(\alpha(2,2) + \alpha(2,2)), \\ \beta(1,0) &= \lambda_0\alpha(2,0) + \lambda_1\alpha(2,1) = \frac{1}{\sqrt{2}}(\alpha(2,1) - \alpha(2,0)), \\ \beta(1,1) &= \lambda_0\alpha(2,2) + \lambda_1\alpha(2,3) = \frac{1}{\sqrt{2}}(\alpha(2,3) - \alpha(2,2)), \end{aligned}$$

and

$$\begin{aligned} \alpha(0,0) &= h_0\alpha(1,0) + h_1\alpha(1,1) = \frac{1}{\sqrt{2}}(\alpha(1,0) + \alpha(1,1)), \\ \beta(0,0) &= \lambda_0\alpha(1,0) + \lambda_1\alpha(1,1) = \frac{1}{\sqrt{2}}(\alpha(1,1) - \alpha(1,0)), \end{aligned}$$

where $h_0 = h_1 = 1/\sqrt{2}$ and $\lambda_0 = -h_1, \lambda_1 = h_0$.

In fact any algorithm could lead to a sign inversion for the vector $\mathbf{b}$ since the mother wavelet is not uniquely defined, see Chapter 5.

Taking the level $j = 1$ gives

$$\begin{aligned} \mathbf{a} &= (\alpha(1,0), \alpha(1,1)) &= (0, 1/\sqrt{2}) \\ \mathbf{b} &= (\beta(1,0), \beta(1,1)) &= (0, 0). \end{aligned}$$

The inverse wavelet transform is obtained via the command

$$\texttt{invfwt}(\texttt{a}, \texttt{b}, \texttt{m}, \texttt{l}, \texttt{h})$$

or alternatively by

$$\texttt{invdwt}(\texttt{y}, \texttt{l}, \texttt{h})$$

Here the entries $\texttt{a}, \texttt{b}$ are the coefficients as above, the entry $\texttt{m} = 2^K$ denotes the length of the input vector, $\texttt{l} = 2^j$ is the number of father wavelets on the input level j.

The thresholding may be done via hard or soft-thresholding, i.e. by transfering the wavelet coefficients through the functions given in (10.13) and (10.14).

Translation invariant wavelet transform

The translation invariant wavelet transform is calculated via the command

$$\texttt{ti} = \texttt{fwtin}(\texttt{x}, \texttt{d}, \texttt{h})$$

where $\texttt{ti}$ is the TI Table, $\texttt{x}$ the input vector as before, $\texttt{d} = j_0$. Note that $\texttt{l} = 2^d$ is the number of father wavelets on the initial level j_0.

The variable $\texttt{h}$ denotes as before the coefficient vector (e.g. symmlet7 for the coefficients of Table 12.2). The inverse transform is called via

$$\texttt{xs} = \texttt{invfwtin}(\texttt{ti}, \texttt{h}).$$

Appendix A

Tables

A.1 Wavelet Coefficients

This table presents the wavelet coefficients for

$$D4, \ldots, D20; S4, S5, \ldots, S10; C1, C2, \ldots, C5,$$

see the description for coefficient extraction in Section 12.6.

1	0.482963	75	0.133197	149	-0.140047	223	-0.00182321
2	0.836516	76	-0.293274	150	-0.107808	224	-0.000720549
3	0.224144	77	-0.0968408	151	0.00401024	225	-0.00379351
4	-0.12941	78	0.148541	152	0.0102682	226	0.0077826
5	0.332671	79	0.0307257	153	0.00188995	227	0.0234527
6	0.806892	80	-0.0676328	154	-0.000302921	228	-0.0657719
7	0.459878	81	0.000250947	155	-0.0149523	229	-0.0611234
8	-0.135011	82	0.0223617	156	0.00380875	230	0.405177
9	-0.0854423	83	-0.0047232	157	0.0491372	231	0.793777
10	0.0352263	84	-0.0042815	158	-0.027219	232	0.428483
11	0.230378	85	0.00184765	159	-0.0519458	233	-0.0717998
12	0.714847	86	0.000230386	160	0.364442	234	-0.0823019
13	0.630881	87	-0.000251963	161	0.777186	235	0.034555
14	-0.0279838	88	3.934732e-005	162	0.48136	236	0.0158805
15	-0.187035	89	0.0266701	163	-0.0612734	237	-0.00900798
16	0.0308414	90	0.188177	164	-0.143294	238	-0.00257452
17	0.032883	91	0.527201	165	0.00760749	239	0.00111752
18	-0.0105974	92	0.688459	166	0.0316951	240	0.000466217

19	0.160102	93	0.281172	167	-0.000542132	241	-7.09833e-005
20	0.603829	94	-0.249846	168	-0.00338242	242	-3.459977e-005
21	0.724309	95	-0.195946	169	0.00106949	243	0.000892314
22	0.138428	96	0.127369	170	-0.000473154	244	-0.00162949
23	-0.242295	97	0.0930574	171	-0.0102641	245	-0.00734617
24	-0.03224	98	-0.0713941	172	0.00885927	246	0.0160689
25	0.0775715	99	-0.0294575	173	0.0620778	247	0.0266823
26	-0.00624149	100	0.0332127	174	-0.0182338	248	-0.0812667
27	-0.0125808	101	0.00360655	175	-0.191551	249	-0.0560773
28	0.00333573	102	-0.0107332	176	0.0324441	250	0.415308
29	0.111541	103	0.00139535	177	0.617338	251	0.782239
30	0.494624	104	0.00199241	178	0.717897	252	0.434386
31	0.751134	105	-0.000685857	179	0.238761	253	-0.0666275
32	0.31525	106	-0.000116467	180	-0.054569	254	-0.0962204
33	-0.226265	107	9.358867e-005	181	0.000583463	255	0.0393344
34	-0.129767	108	-1.32642e-005	182	0.0302249	256	0.0250823
35	0.0975016	109	-0.0757657	183	-0.0115282	257	-0.0152117
36	0.0275229	110	-0.0296355	184	-0.013272	258	-0.00565829
37	-0.031582	111	0.497619	185	0.000619781	259	0.00375144
38	0.000553842	112	0.803739	186	0.00140092	260	0.00126656
39	0.00477726	113	0.297858	187	0.00077016	261	-0.000589021
40	-0.0010773	114	-0.0992195	188	9.563267e-005	262	-0.000259975
41	0.0778521	115	-0.012604	189	-0.00864133	263	6.233903e-005
42	0.396539	116	0.0322231	190	-0.00146538	264	3.122988e-005
43	0.729132	117	0.0273331	191	0.0459272	265	-3.25968e-006
44	0.469782	118	0.0295195	192	0.0116099	266	-1.784985e-006
45	-0.143906	119	-0.0391342	193	-0.159494	267	-0.000212081
46	-0.224036	120	0.199398	194	-0.0708805	268	0.00035859
47	0.0713092	121	0.723408	195	0.471691	269	0.00217824
48	0.0806126	122	0.633979	196	0.76951	270	-0.00415936
49	-0.0380299	123	0.0166021	197	0.383827	271	-0.0101311
50	-0.0165745	124	-0.175328	198	-0.0355367	272	0.0234082
51	0.012551	125	-0.0211018	199	-0.0319901	273	0.028168
52	0.00042957	126	0.0195389	200	0.049995	274	-0.09192
53	-0.0018016	127	0.0154041	201	0.00576491	275	-0.0520432
54	0.00035371	128	0.00349071	202	-0.0203549	276	0.421566
55	0.0544158	129	-0.11799	203	-0.000804359	277	0.77429
56	0.312872	130	-0.0483117	204	0.00459317	278	0.437992
57	0.675631	131	0.491055	205	5.703608e-005	279	-0.062036
58	0.585355	132	0.787641	206	-0.000459329	280	-0.105574
59	-0.0158291	133	0.337929	207	-0.0727326	281	0.0412892

60	-0.284016	134	-0.0726375	208	0.337898	282	0.0326836
61	0.000472485	135	-0.0210603	209	0.852572	283	-0.0197618
62	0.128747	136	0.0447249	210	0.384865	284	-0.00916423
63	-0.0173693	137	0.00176771	211	-0.072733	285	0.00676419
64	-0.0440883	138	-0.00780071	212	-0.0156557	286	0.00243337
65	0.013981	139	0.00268181	213	0.0163873	287	-0.00166286
66	0.00874609	140	-0.00104738	214	-0.0414649	288	-0.000638131
67	-0.00487035	141	-0.0126363	215	-0.0673726	289	0.00030226
68	0.000039174	142	0.0305155	216	0.38611	290	0.000140541
69	0.000675449	143	0.0678927	217	0.812724	291	-4.134043e-005
70	-0.000117477	144	-0.0495528	218	0.417005	292	-2.131503e-005
71	0.0380779	145	0.0174413	219	-0.0764886	293	3.734655e-006
72	0.243835	146	0.536102	220	-0.0594344	294	2.063762e-006
73	0.604823	147	0.767764	221	0.0236802	295	-1.674429e-007
74	0.657288	148	0.28863	222	0.00561143	296	-9.517657e-008

Table A.1: The 296 coefficients for the wavelet construction.

A.2

1	0	0	12	117	126
2	1	4	13	127	138
3	5	10	14	139	152
4	11	18	15	153	168
5	19	28	16	169	186
6	29	40	17	187	206
7	41	54	18	207	212
8	55	70	19	213	224
9	71	88	20	225	242
10	89	108	21	243	266
11	109	116	22	267	296

Table A.2: The indices for the selected wavelets. The first column indicates the wavelet number, the second the lower index, the third the upper index.

Appendix B

Software Availability

For questions concerning the availability of new releases of XploRe, contact
`xplore@netcologne.de`

    or

GfKI – Gesellschaft für Kommunikation und Information
Mauritiussteinweg 2
D-50676 Köln
GERMANY
FAX:+49 22 1923 3906

    There exists a mailing list for discussion of software problems. Mail to

`stat@wiwi.hu-berlin.de`

for subscribing or unsubscribing to the mailing list. After subscribing, send your mail to:

`xplore@wiwi.hu-berlin.de`

    The XploRe programs that produced the figures in this text are freely distributed. The whole set of programs is available via internet by contacting

`http://wotan.wiwi.hu-berlin.de`

You may be interested in trying the Java interface of XploRe

    All algorithms in this book are freely available. They can be found under the above `http` adress under

`http://wotan.wiwi.hu-berlin.de`

Putting the algorithm `hkpt103.xpl` into the Java interface results in a graph corresponding to a picture 10.3 in this text. The other graphes may be recalculated correspondingly.

Appendix C

Bernstein and Rosenthal inequalities

The aim of this appendix is to give a simple proof of both Bernstein and Rosenthal inequalities. For a deeper insight into the field of general moments or exponential inequalities we refer to Petrov (1995), Pollard (1984), Hall & Heyde (1980) (for the case of martingales), Ledoux & Talagrand (1991) for more general isoperimetric and concentration of measure inequalities. The proof is based on the following lemma which is a special case of concentration of measure results.

LEMMA C.1 *Let $X_1, \ldots, X_n$ be independent random variables such that $X_i \leq M, E(X_i) \leq 0, b_n^2 = \sum\limits_{i=1}^{n} E(X_i^2)$. Then for any $\lambda \geq 0$,*

$$P(\sum_{i=1}^{n} X_i \geq \lambda) \leq \exp\left(-\frac{b_n^2}{M^2}\theta\left(\frac{\lambda M}{b_n^2}\right)\right) \tag{C.1}$$

where $\theta(x) = (1 + x)\log(1 + x) - x$.

Proof:

- Consider the function

$$\Phi(x) = \begin{cases} (e^x - 1 - x)/x^2 & , \quad x \neq 0, \\ \frac{1}{2} & , \quad x = 0. \end{cases}$$

241

Clearly $\Phi(x) \geq 0$, $\forall x \in \mathbb{R}^1$, and $\Phi(x)$ is non-decreasing. The last property is easily obtained by observing that the derivative of Φ is $\frac{1}{x^3}(e^x(x-2)+x+2)$, $x \neq 0$, and then proving that $e^x(x-2)+x+2$ has the same sign as x.

- Using the Markov inequality and independence of X'_is we get that, for arbitrary $t > 0, \lambda > 0$,

$$P\left(\sum_{i=1}^{n} X_i > \lambda\right) \leq \exp(-\lambda t)E\left[\exp\left(\sum_{i=1}^{n} tX_i\right)\right]$$

$$= \exp\left\{-\left[\lambda t - \sum_{i=1}^{n} \log E(e^{tX_i})\right]\right\}.$$

Next,

$$\log E(e^{tX_i}) = \log\left(E(e^{tX_i} - 1 - tX_i + 1 + tX_i)\right)$$
$$\leq \log\left(E(e^{tX_i} - 1 - tX_i) + 1\right)$$
$$= \log\left(1 + E(\Phi(tX_i)t^2 X_i^2)\right),$$

where we used the inequality $E(X_i) \leq 0$. Thus, since $\log(1+u) \leq u$ for $u \geq 0$, we get

$$\log E(e^{tX_i}) \leq E\left(\Phi(tX_i)t^2 X_i^2\right) \leq \Phi(tM)t^2 E(X_i^2),$$

using the monotonicity of the function Φ. Then it follows:

$$P\left(\sum_{i=1}^{n} X_i > \lambda\right) \leq \exp\left\{-[\lambda t - b_n^2 t^2 \Phi(tM)]\right\}$$

$$= \exp\left\{-\frac{b_n^2}{M^2}\left[\frac{\lambda M^2}{b_n^2}t - (e^{tM} - 1 - tM)\right]\right\}.$$

As $t > 0$ can be arbitrary, we optimize this inequality by taking t such that

$$\frac{\lambda M^2}{b_n^2} - Me^{tM} + M = 0 \Leftrightarrow t = \frac{1}{M}\log(1 + \frac{\lambda M}{b_n^2}),$$

wich gives the result.

$\square$

We now prove the following result known as Bernstein's inequality (see Petrov (1995), Pollard (1984) for complete bibliography).

THEOREM C.1 *Under the assumptions of Lemma C.1, for any $\lambda > 0$,*

$$P\left(\sum_{i=1}^{n} X_i > \lambda\right) \leq \exp\left(-\frac{\lambda^2}{2(b_n^2 + \frac{\lambda M}{3})}\right).$$

Proof:
It suffices to show that in inequality (C.1) one can replace the function $\theta(x)$ by the function

$$h(x) = \frac{3}{2}\frac{x^2}{x+3}.$$

Hence, we have to prove that

$$\theta(x) - h(x) \geq 0, \forall x \geq 0.$$

This is easily done by observing that $\theta(0) = h(0), \theta'(0) = h'(0)$ and $\theta''(0) \geq h''(0), \forall x \geq 0.$ □

The following Corollary is a direct consequence of Theorem C.1.

COROLLARY C.1 *(i) If X_i are independent random variables, $|X_i| \leq M, E(X_i) = 0$, then*

$$P\left(\left|\sum_{i=1}^{n} X_i\right| \geq \lambda\right) \leq 2\exp\left(-\frac{\lambda^2}{2(b_n^2 + \frac{\lambda M}{3})}\right), \quad \forall \lambda \geq 0.$$

(ii) If X_i are i.i.d , $|X_i| \leq M, E(X_i) = 0, E(X_i^2) = \sigma^2$, then

$$P\left(\frac{1}{n}\left|\sum_{i=1}^{n} X_i\right| \geq v\right) \leq 2\exp\left(-\frac{nv^2}{2(\sigma^2 + \frac{vM}{3})}\right), \forall v \geq 0.$$

Let us now prove the following result known as Rosenthal's inequality (Rosenthal (1970)).

THEOREM C.2 *Let $p \geq 2$ and let $(X_1, \ldots, X_n)$ be independent random variables such that $E(X_i) = 0, E(|X_i|^p) < \infty$. Then there exists $C(p)$ such that*

$$E\left(\left|\sum_{i=1}^{n} X_i\right|^p\right) \leq C(p) \left\{\sum_{i=1}^{n} E\left(|X_i|^p\right) + \left(\sum_{i=1}^{n} E(X_i^2)\right)^{p/2}\right\}. \qquad (C.2)$$

REMARK C.1 *This inequality is an extension of the classical convexity inequalities, true for $0 < p \leq 2$:*

$$E\left(\left|\sum_{i=1}^{n} X_i\right|^p\right) \leq \left(E\left(\left|\sum_{i=1}^{n} X_i\right|^2\right)\right)^{p/2} = \left(\sum_{i=1}^{n} E(X_i)^2\right)^{p/2}.$$

Proof:
We use again Lemma C.1, but this time we replace $\theta(x)$ by $x \log(1 + x) - x$ which is obviously smaller than $\theta(x)$ for any $x \geq 0$. Let us fix an arbitrary $y \geq 0$ and consider the random variables $Y_i = X_i I\{X_i \leq y\}$. We have $E(Y_i) \leq E(X_i) = 0, Y_i \leq y$, and

$$B_n^2 = \sum_{i=1}^{n} E(X_i^2) \geq \sum_{i=1}^{n} E(Y_i^2) = b_n^2.$$

It follows from Lemma C.1 that

$$\begin{aligned}
P\left(\sum_{i=1}^{n} Y_i \geq x\right) &\leq \exp\left\{-\frac{b_n^2}{y^2}\theta\left(\frac{xy}{b_n^2}\right)\right\} \\
&\leq \exp\left[-\frac{b_n^2}{y^2}\left\{\frac{xy}{b_n^2}\log\left(1 + \frac{xy}{b_n^2}\right) - \frac{xy}{b_n^2}\right\}\right] \\
&\leq \exp\left[-\frac{x}{y}\left\{\log\left(1 + \frac{xy}{B_n^2}\right) - 1\right\}\right], \quad \forall x > 0.
\end{aligned}$$

Using this inequality we get, for any $x > 0$,

$$\begin{aligned}
P\left(\sum_{i=1}^{n} X_i > x\right) &\leq P\left(\sum_{i=1}^{n} Y_i > x, X_1 \leq y, \ldots, X_n \leq y\right) + P\left(\max_{1 \leq i \leq n} X_i > y\right) \\
&\leq P\left(\sum_{i=1}^{n} Y_i > x\right) + \sum_{i=1}^{n} P(X_i > y) \\
&\leq \sum_{i=1}^{n} P(X_i > y) + \exp\left(-\frac{x}{y}\left\{\log\left(1 + \frac{xy}{B_n^2}\right) - 1\right\}\right). \quad (C.3)
\end{aligned}$$

Quite similarly one obtains

$$P\left(\sum_{i=1}^{n}(-X_i) > x\right) \leq \sum_{i=1}^{n} P(-X_i > y)$$

$$+ \exp\left(-\frac{x}{y}\left\{\log\left(1 + \frac{xy}{B_n^2}\right) - 1\right\}\right). \qquad (C.4)$$

Combining (C.3) and (C.4), and putting $y = x/\tau$, $\tau > 0$, we find

$$P\left(\left|\sum_{i=1}^{n} X_i\right| > x\right) \leq \sum_{i=1}^{n} P(\tau|X_i| > x) + 2\exp\left(-\tau\left\{\log\left(1 + \frac{x^2}{\tau B_n^2}\right) - 1\right\}\right).$$

Now, for $p > 1$,

$$E\left(\left|\sum_{i=1}^{n} X_i\right|^p\right) = \int_0^\infty px^{p-1} P\left(\left|\sum_{i=1}^{n} X_i\right| > x\right) dx$$

$$\leq \sum_{i=1}^{n} \int_0^\infty px^{p-1} P(\tau|X_i| > x) dx$$

$$+ 2p \int_0^\infty x^{p-1} \exp\left(-\tau\left\{\log\left(1 + \frac{x^2}{\tau B_n^2}\right) - 1\right\}\right) dx$$

$$\leq \sum_{i=1}^{n} E\left(|\tau X_i|^p\right)$$

$$+ p(\tau B_n^2)^{p/2} e^\tau \int_0^\infty t^{\frac{p-2}{2}} (1+t)^{-\tau} dt, \qquad (C.5)$$

where we made the change of the variables $t = \frac{x^2}{\tau B_n^2}$. To end the proof it remains to choose τ such that the integral of the RHS is convergent i.e. $\tau > p/2$. Under this choice of τ inequality (C.5) entails (C.2) with

$$C(p) = \max\{\tau^p, p\tau^{p/2} e^\tau \int_0^\infty t^{\frac{p-2}{2}} (1+t)^{-\tau} dt\}.$$

□

Appendix D

A Lemma on the Riesz basis

We prove that if $\{g(\cdot - k), k \in \mathbb{Z}\}$ is a Riesz basis, then (6.1) is satisfied. Thus, we complete the proof of Proposition 6.1. Note that is $\{g(\cdot - k), k \in \mathbb{Z}\}$ a Riesz basis, then the following property is true.

For every trigonometric polynomial $m(\xi) = \sum\limits_{-N}^{+N} a_k e^{-ik\xi}$ we have:

$$A\frac{1}{2\pi} \int_0^{2\pi} |m(\xi)|^2 d\xi \le \frac{1}{2\pi} \int_0^{2\pi} \Gamma(\xi)|m(\xi)|^2 d\xi \le B\frac{1}{2\pi} \int_0^{2\pi} |m(\xi)|^2 d\xi \quad \text{(D.1)}$$

Let us prove that this implies $A \le \Gamma(\xi) \le B$ a.e. If we introduce the following Fejer kernel:

$$K_N(\xi) = \frac{1}{N} \sum_{k=-N}^{N} \left(1 - \frac{|k|}{N}\right) e^{ik\xi},$$

it is well known (see for instance Katznelson (1976),p.11) that,

$$K_N * \Gamma(\xi_0) = \frac{1}{2\pi} \int_0^{2\pi} K_N(\xi_0 - \xi)\Gamma(\xi)d\xi$$

converges in L_1 to $\Gamma(\xi_0)$ as $N \to \infty$. So there exists a subsequence N' such that $K_{N'} * \Gamma(\cdot) \to \Gamma(\cdot)$ a.e., as $N' \to \infty$. (in fact this result is also true without taking a subsequence but is much more difficult to prove.) Recall that

$$K_N(\xi) = \frac{1}{N} \left(\frac{\sin \frac{N\xi}{2}}{\sin \frac{\xi}{2}}\right)^2,$$

and that for

$$D_N(\xi) = \sum_{k=-N}^{N} e^{ik\xi} = \frac{\sin\frac{(2N+1)\xi}{2}}{\sin\frac{\xi}{2}}$$

we have

$$K_{2N+1}(\xi) = \left|\frac{1}{\sqrt{2N+1}} D_N(\xi)\right|^2.$$

As

$$\frac{1}{2\pi}\int_0^{2\pi} K_{2N+1}(\xi)d\xi = 1$$

using (D.1) we deduce

$$A \le \frac{1}{2\pi}\int_0^{2\pi} K_{2N+1}(\xi_0 - \xi)\Gamma(\xi)d\xi \le B$$

and using the a.e. convergence of the subsequence $K_{2N'+1}$, we deduce (6.1).

Bibliography

Abramovich, F. & Benjamini, Y. (1996). Adaptive thresholding of wavelet coefficients, *Computational Statistics and Data Analysis* **22**: 351–361.

Adams, R. (1975). *Sobolev Spaces*, Academic Press, New York.

Antoniadis, A. (1994). Smoothing noisy data with tapered coiflet series, *Technical Report RR 993-M*, University of Grenoble.

Antoniadis, A., Grégoire, G. & McKeague, I. (1994). Wavelet methods for curve estimation, *Journal of the American Statistical Association* **89**: 1340–1353.

Antoniadis, A. & Oppenheim, G. (eds) (1995). *Wavelets and Statistics*, Vol. 103 of *Lecture Notes in Statistics*, Springer, Heidelberg.

Assouad, P. (1983). Deux remarques sur l'estimation, *Comptes Rendus Acad. Sci.Paris (A)* **296**: 1021–1024.

Auscher, P. (1992). Solution of two problems on wavelets, *Preprint*, IRMAR, Univ. Rennes I.

Bergh, J. & Löfström, J. (1976). *Interpolation spaces - An Introduction*, Springer Verlag, New York.

Besov, O. V., Il'in, V. L. & Nikol'skii, S. M. (1978). *Integral Representations of Functions and Embedding Theorems.*, J. Wiley, New York.

Beylkin, G., Coifman, R. R. & Rokhlin, V. (1991). Fast wavelet transforms and numerical algorithms, *Comm. Pure and Appl. Math.* **44**: 141–183.

Birgé, L. (1983). Approximation dans les éspaces metriques et théorie de l'estimation, *Zeitschrift für Wahrscheinlichkeitstheorie und verwandte Gebiete* **65**: 181–237.

Birgé, L. & Massart, P. (1997). From model selection to adaptive estimation, *in* D. Pollard (ed.), *Festschrift for L. Le Cam*, Springer, pp. 55–88.

Black, F. & Scholes, M. (1973). The pricing of options and corporate liabilities, *Journal of Political Economy* **81**: 637–654.

Bossaerts, P., Hafner, C. & Härdle, W. (1996). Foreign exchange-rates have surprising volatility, *in* P. Robinson (ed.), *Ted Hannan Memorial Volume*, Springer Verlag.

Bretagnolle, J. & Huber, C. (1979). Estimation des densités: risque minimax, *Z. Wahrscheinlichkeitstheorie und Verwandte Gebiete* **47**: 119–137.

Brown, L.-D. & Low, M. L. (1996). Asymptotic equivalence of nonparametric regression and white noise, *Annals of Statistics* **24**: 2384–2398.

Bruce, A. & Gao, H.-Y. (1996a). *Applied Wavelet Analysis with S-Plus*, Springer Verlag, Heidelberg, New York.

Bruce, A. & Gao, H.-Y. (1996b). Understanding waveshrink: variance and bias estimation, *Biometrika* **83**: 727–745.

Burke-Hubbard, B. (1995). *Ondes et ondelettes*, Pour la science, Paris.

Centsov, N. N. (1962). Evaluation of an unknown distribution density from observations, *Soviet Math. Dokl.* **3**: 1599–1562.

Chui, C. (1992a). *An Introduction to Wavelets*, Academic Press, Boston.

Chui, C. (1992b). *Wavelets: a Tutorial in Theory and Applications*, Academic Press, Boston.

Cohen, A., Daubechies, I. & Vial, P. (1993). Wavelets on the interval and fast wavelet transform, *Journal of Applied and Computational Harmonic Analysis* **1**: 54–81.

Cohen, A. & Ryan, R. (1995). *Wavelets and Multiscale Signal Processing*, Chapman & Hall.

Coifman, R. R. & Donoho, D. (1995). Translation-invariant de-noising, *in* Antoniadis & Oppenheim (1995), pp. 125–150.

Dahlhaus, R. (1997). Fitting time series models to nonstationary processes, *Annals of Statistics* **25**: 1–37.

Daubechies, I. (1988). Orthonormal bases of compactly supported wavelets, *Comm. Pure and Appl. Math.* **41**: 909–996.

Daubechies, I. (1992). *Ten Lectures on Wavelets*, SIAM, Philadelphia.

Delyon, B. & Juditsky, A. (1996a). On minimax wavelet estimators, *Journal of Applied and Computational Harmonic Analysis* **3**: 215–228.

Delyon, B. & Juditsky, A. (1996b). On the computation of wavelet coefficients, *Technical report*, IRSA/INRIA, Rennes.

DeVore, R. A. & Lorentz, G. (1993). *Constructive Approximation*, Springer-Verlag, New York.

Donoho, D. (1992a). De-noising via soft-thresholding, *Technical report 409*, Dept. of Statistics, Stanford University.

Donoho, D. (1992b). Interpolating wavelet transforms, *Technical report 408*, Dept. of Statistics, Stanford University.

Donoho, D. (1993). Smooth wavelet decompositions with blocky coefficient kernels, *Technical report*, Dept. of Statistics, Stanford University.

Donoho, D. (1994). Statistical estimation and optimal recovery, *Annals of Statistics* **22**: 238–270.

Donoho, D. (1995). Nonlinear solutions of linear inverse problems by wavelet-vaguelette decomposition, *Journal of Applied and Computational Harmonic Analysis* **2**: 101–126.

Donoho, D. & Johnstone, I. (1991). Minimax estimation via wavelet shrinkage, *Tech. Report, Stanford University*.

Donoho, D. & Johnstone, I. (1994a). Ideal spatial adaptation by wavelet shrinkage, *Biometrika* **81**: 425–455.

Donoho, D. & Johnstone, I. (1994b). Minimax risk over l_p-balls for l_p-error, *Probabiliy Theory and Related Fields* **99**: 277–303.

Donoho, D. & Johnstone, I. (1995). Adapting to unknown smoothness via wavelet shrinkage, *Journal of the American Statistical Association* **90**: 1200–1224.

Donoho, D. & Johnstone, I. (1996). Neoclassical minimax problems, thresholding and adaptive function estimation, *Bernoulli* **2**: 39–62.

Donoho, D., Johnstone, I., Kerkyacharian, G. & Picard, D. (1995). Wavelet shrinkage: Asymptopia?, *Journal of the Royal Statistical Society, Series B* **57**: 301–369.

Donoho, D., Johnstone, I., Kerkyacharian, G. & Picard, D. (1996). Density estimation by wavelet thresholding, *Annals of Statistics* **24**: 508–539.

Donoho, D., Johnstone, I., Kerkyacharian, G. & Picard, D. (1997). Universal near minimaxity of wavelet shrinkage, *in* D. Pollard (ed.), *Festschrift for L. Le Cam*, Springer, N.Y. e.a., pp. 183–218.

Donoho, D., Mallat, S. G. & von Sachs, R. (1996). Estimating covariances of locally stationary processes: Consistency of best basis methods, *Technical report*, University of Berkeley.

Doukhan, P. (1988). Formes de Töeplitz associées à une analyse multiéchelle, *Comptes Rendus Acad. Sci.Paris (A)* **306**: 663–666.

Doukhan, P. & Leon, J. (1990). Déviation quadratique d'estimateurs d'une densité par projection orthogonale, *Comptes Rendus Acad. Sci. Paris, (A)* **310**: 425–430.

Efroimovich, S. (1985). Nonparametric estimation of a density with unknown smoothness, *Theory of Probability and its Applications* **30**: 524–534.

Efroimovich, S. & Pinsker, M. (1981). Estimation of square-integrable density on the basis of a sequence of observations, *Problems of Information Transmission* **17**: 182–195.

Fama, E. F. (1976). *Foundations of Finance*, Basil Blackwell, Oxford.

Fan, J. (1994). Test of significance based on wavelet thresholding and Neyman's truncation. Preprint.

Fix, G. & Strang, G. (1969). A Fourier analysis of the finite element method, *Stud. Appl. Math.* **48**: 265–273.

Foufoula-Georgiou, E. & Kumar, P. (eds) (1994). *Wavelets in Geophysics*, Academic Press, Boston/London/Sydney.

Gao, H.-Y. (1993a). Choice of thresholds for wavelet estimation of the log spectrum. Preprint 430. Dept. of Stat. Stanford University.

Gao, H.-Y. (1993b). Wavelet estimation of spectral densities in time series analysis. PhD Dissertation. University of California, Berkeley.

Gasser, T., Stroka, L. & Jennen-Steinmetz, C. (1986). Residual variance and residual pattern in nonlinear regression, *Biometrika* **73**: 625–633.

Genon-Catalot, V., Laredo, C. & Picard, D. (1992). Nonparametric estimation of the variance of a diffusion by wavelet methods, *Scand. Journal of Statistics* **19**: 319–335.

Ghysels, E., Gourieroux, C. & Jasiak, J. (1995). Trading patterns, time deformation and stochastic volatility in foreign exchange markets, *Discussion paper*, CREST, Paris.

Gourieroux, C. (1992). *Modèles ARCH et Applications Financières*, Economica, Paris.

Hall, P. & Heyde, C. C. (1980). *Martingale Limit Theory and its Applications*, Acad. Press, New York.

Hall, P., Kerkyacharian, G. & Picard, D. (1996a). Adaptive minimax optimality of block thresholded wavelet estimators, *Statistica Sinica*. Submitted.

Hall, P., Kerkyacharian, G. & Picard, D. (1996b). Note on the wavelet oracle, *Technical report*, Aust. Nat. University, Canberra.

Hall, P., Kerkyacharian, G. & Picard, D. (1996c). On block thresholding for curve estimators using kernel and wavelet methods. Submitted.

Hall, P., McKay, I. & Turlach, B. A. (1996). Performance of wavelet methods for functions with many discontinuities, *Annals of Statistics* **24**: 2462–2476.

Hall, P. & Patil, P. (1995a). Formulae for mean integrated squared error of nonlinear wavelet-based density estimators, *Annals of Statistics* **23**: 905–928.

Hall, P. & Patil, P. (1995b). On wavelet methods for estimating smooth functions, *Bernoulli* **1**: 41–58.

Hall, P. & Patil, P. (1996a). Effect of threshold rules on performance of wavelet-based curve estimators, *Statistica Sinica* **6**: 331–345.

Hall, P. & Patil, P. (1996b). On the choice of smoothing parameter, threshold and truncation in nonparametric regression by nonlinear wavelet methods, *Journal of the Royal Statistical Society, Series B* **58**: 361–377.

Hall, P. & Turlach, B. A. (1995). Interpolation methods for nonlinear wavelet regression with irregularly spaced design. Preprint.

Härdle, W. (1990). *Applied Nonparametric Regression*, Cambridge University Press, Cambridge.

Härdle, W., Klinke, S. & Turlach, B. A. (1995). *XploRe - an Interactive Statistical Computing Environment*, Springer, Heidelberg.

Härdle, W. & Scott, D. W. (1992). Smoothing by weighted averaging of rounded points, *Computational Statistics* **7**: 97–128.

Hildenbrand, W. (1994). *Market Demand*, Princeton University Press, Princeton.

Hoffmann, M. (1996). *Méthodes adaptatives pour l'estimation non-paramétrique des coefficients d'une diffusion*, Phd thesis, Université Paris VII.

Holschneider, M. (1995). *Wavelets: an Analysis Tool*, Oxford University Press, Oxford.

Ibragimov, I. A. & Hasminskii, R. Z. (1980). On nonparametric estimation of regression, *Soviet Math. Dokl.* **21**: 810–814.

Ibragimov, I. A. & Hasminskii, R. Z. (1981). *Statistical Estimation: Asymptotic Theory*, Springer, New York.

Johnstone, I. (1994). Minimax Bayes, asymptotic minimax and sparse wavelet priors, *in* S.Gupta & J.Berger (eds), *Statistical Decision Theory and Related Topics*, Springer, pp. 303–326.

Johnstone, I., Kerkyacharian, G. & Picard, D. (1992). Estimation d'une densité de probabilité par méthode d'ondelette, *Comptes Rendus Acad. Sci. Paris, (1)* **315**: 211–216.

Johnstone, I. & Silverman, B. W. (1997). Wavelet methods for data with correlated noise, *Journal of the Royal Statistical Society, Series B* **59**: 319–351.

Juditsky, A. (1997). Wavelet estimators: adapting to unknown smoothness, *Mathematical Methods of Statistics* **6**: 1–25.

Kahane, J. P. & Lemarié-Rieusset, P. (1995). *Fourier Series and Wavelets*, Gordon and Breach Science Publishers, Amsterdam.

Kaiser, G. (1995). *A Friendly Guide to Wavelets*, Birkhäuser, Basel.

Katznelson, Y. (1976). *An Introduction to Harmonic Analysis*, Dover, New York.

Kerkyacharian, G. & Picard, D. (1992). Density estimation in Besov spaces, *Statistics and Probability Letters* **13**: 15–24.

Kerkyacharian, G. & Picard, D. (1993). Density estimation by kernel and wavelet methods: optimality of Besov spaces, *Statistics and Probability Letters* **18**: 327–336.

Kerkyacharian, G., Picard, D. & Tribouley, K. (1996). L_p adaptive density estimation, *Bernoulli* **2**: 229–247.

Korostelev, A. P. & Tsybakov, A. B. (1993a). Estimation of the density support and its functionals, *Problems of Information Transmission* **29**: 1–15.

Korostelev, A. P. & Tsybakov, A. B. (1993b). *Minimax Theory of Image Reconstruction*, Springer, New York.

Leadbetter, M. R., Lindgren, G. & Rootzén, H. (1986). *Extremes and Related Properties of Random Sequences and Processes*, Springer, N.Y e.a.

Ledoux, M. & Talagrand, M. (1991). *Probability in Banach Spaces*, Springer, New York.

Lemarié, P. (1991). Fonctions à support compact dans les analyses multi-résolutions, *Revista Mat. Iberoamericana* **7**: 157–182.

Lemarié-Rieusset, P. (1993). Ondelettes généralisées et fonctions d'échelle à support compact, *Revista Mat. Iberoamericana* **9**: 333–371.

Lemarié-Rieusset, P. (1994). Projecteurs invariants, matrices de dilatation, ondelettes et analyses multi-résolutions, *Revista Mat. Iberoamericana* **10**: 283–347.

Lepski, O., Mammen, E. & Spokoiny, V. (1997). Optimal spatial adaptation to inhomogeneous smoothness: an approach based on kernel estimates with variable bandwidth selectors, *Annals of Statistics* **25**: 929–947.

Lepski, O. & Spokoiny, V. (1995). Local adaptation to inhomogeneous smoothness: resolution level, *Mathematical Methods of Statistics* **4**: 239–258.

Lepskii, O. (1990). On a problem of adaptive estimation in gaussian white noise, *Theory Prob. Appl.* **35**: 454–466.

Lepskii, O. (1991). Asymptotically minimax adaptive estimation I: Upper bounds. Optimal adaptive estimates, *Theory Prob. Appl.* **36**: 682–697.

Lepskii, O. (1992). Asymptotically minimax adaptive estimation II: Statistical models without optimal adaptation. Adaptive estimates, *Theory Prob. Appl.* **37**: 433–468.

Lintner, J. (1965). Security prices, risk and maximal gains from diversification, *Journal of Finance* **20**: 587–615.

Mallat, S. G. (1989). A theory for multiresolution signal decomposition: the wavelet representation, *IEEE Transactions on Pattern Analysis and Machine Intelligence* **11**: 674–693.

Marron, J. S., Adak, S., Johnstone, I., Neumann, M. & Patil, P. (1995). Exact risk analysis of wavelet regression. Manuscript.

Marron, J. S. & Tsybakov, A. B. (1995). Visual error criteria for qualitative smoothing, *Journal of the American Statistical Association* **90**: 499–507.

Meyer, Y. (1990). *Ondelettes et opérateurs*, Hermann, Paris.

Meyer, Y. (1991). Ondelettes sur l'intervalle, *Rev. Mat. Iberoamericana* **7**: 115–133.

Meyer, Y. (1993). *Wavelets: Algorithms and Applications*, SIAM, Philadelphia.

Misiti, M., Misiti, Y., Oppenheim, G. & Poggi, J. (1996). *Wavelet TOOLBOX*, The MathWorks Inc., Natick, MA.

Moulin, P. (1993). Wavelet thresholding techniques for power spectrum estimation, *IEEE. Trans. Signal Processing* **42**: 3126–3136.

Nason, G. (1996). Wavelet shrinkage using cross-validation, *Journal of the Royal Statistical Society, Series B* **58**: 463–479.

Nason, G. & Silverman, B. W. (1994). The discrete wavelet transform in S, *Journal of Computational and Graphical Statistics* **3**: 163–191.

Nemirovskii, A. S. (1986). Nonparametric estimation of smooth regression functions, *Journal of Computer and System Sciences* **23**(6): 1–11.

Nemirovskii, A. S., Polyak, B. T. & Tsybakov, A. B. (1983). Estimators of maximum likelihood type for nonparametric regression, *Soviet Math. Dokl.* **28**: 788–92.

Nemirovskii, A. S., Polyak, B. T. & Tsybakov, A. B. (1985). Rate of convergence of nonparametric estimators of maximum likelihood type, *Problems of Information Transmission* **21**: 258–272.

Neumann, M. (1996a). Multivariate wavelet thresholding: a remedy against the curse of dimensionality ? Preprint 239. Weierstrass Inst. of Applied Analysis and Stochastics, Berlin.

Neumann, M. (1996b). Spectral density estimation via nonlinear wavelet methods for stationary non-gaussian time series, *Journal of Time Series Analysis* **17**: 601–633.

Neumann, M. & Spokoiny, V. (1995). On the efficiency of wavelet estimators under arbitrary error distributions, *Mathematical Methods of Statistics* **4**: 137–166.

Neumann, M. & von Sachs, R. (1995). Wavelet thresholding: beyond the Gaussian iid situation, *in* Antoniadis & Oppenheim (1995), pp. 301–329.

Neumann, M. & von Sachs, R. (1997). Wavelet thresholding in anisotropic function classes and application to adaptive estimation of evolutionary spectra, *Annals of Statistics* **25**: 38–76.

Nikol'skii, S. M. (1975). *Approximation of Functions of Several Variables and Imbedding Theorems*, Springer, New York.

Nussbaum, M. (1985). Spline smoothing in regression models and asymptotic efficiency in L_2, *Annals of Statistics* **13**: 984–97.

Nussbaum, M. (1996). Asymptotic equivalence of density estimation and gaussian white noise, *Annals of Statistics* **24**: 2399–2430.

Ogden, T. (1997). *Essential Wavelets for Statistical Applications and Data Analysis*, Birkhäuser, Basel.

Ogden, T. & Parzen, E. (1996). Data dependent wavelet thresholding in nonparametric regression with change point applications, *Computational Statistics and Data Analysis* **22**: 53–70.

Oppenheim, A. & Schafer, R. (1975). *Digital Signal Processing*, Prentice-Hall, New York.

Papoulis, G. (1977). *Signal Analysis*, McGraw Hill.

Park, B. V. & Turlach, B. A. (1992). Practical performance of several data driven bandwidth selectors, *Computational Statistics* **7**: 251–270.

Peetre, J. (1975). New thoughts on Besov spaces, vol. 1, *Technical report*, Duke University, Durham, NC.

Pesquet, J. C., Krim, H. & Carfantan, H. (1994). Time invariant orthogonal wavelet representation. Submitted for publication.

Petrov, V. V. (1995). *Limit Theorems of Probability Theory*, Clarendon Press, Oxford.

Pinsker, M. (1980). Optimal filtering of square integrable signals in gaussian white noise, *Problems of Information Transmission* **16**: 120–133.

Pollard, D. (1984). *Convergence of Stochastic Processes*, Springer, New York.

Raimondo, M. (1996). *Modelles en ruptures*, Phd thesis, Université Paris VII.

Rioul, O. & Vetterli, M. (1991). Wavelets and signal processing, *IEEE Signal Processing Magazine* **8(4)**: 14–38.

Rosenthal, H. P. (1970). On the subspaces of $L^p (p > 2)$ spanned by sequences of independent random variables, *Israel Journal of Mathematics* **8**: 273–303.

Sharpe, W. (1964). Capital asset prices: a theory of market equilibrium under conditions of risk, *Journal of Finance* **19**: 425–442.

Silverman, B. W. (1986). *Density Estimation for Statistics and Data Analysis*, Chapman and Hall, London.

Spokoiny, V. (1996). Adaptive hypothesis testing using wavelets, *Annals of Statistics* **25**: 2477–2498.

Stein, C. M. (1981). Estimation of the mean of a multivariate normal distribution, *Annals of Statistics* **9**: 1135–1151.

Stein, E. & Weiss, G. (1971). *Introduction to Fourier Analysis on Euclidean Spaces*, Princeton University Press, Princeton.

Stone, C. J. (1980). Optimal rates of convergence for nonparametric estimators, *Annals of Statistics* **8**: 1348–60.

Stone, C. J. (1982). Optimal global rates of convergence for nonparametric regression, *Annals of Statistics* **10**: 1040–1053.

Strang, G. & Nguyen, T. (1996). *Wavelets and Filter Banks*, Wellesley-Cambridge Press, Wellesley, MA.

Tribouley, K. (1995). Practical estimation of multivariate densities using wavelet methods, *Statistica Neerlandica* **49**: 41–62.

Tribouley, K. & Viennet, G. (1998). L_p adaptive estimation of the density in a β-mixing framework., *Ann. de l'Institut H. Poincaré, to appear*.

Triebel, H. (1992). *Theory of Function Spaces II*, Birkhäuser Verlag, Basel.

Tsybakov, A. B. (1995). Pointwise and sup-norm adaptive signal estimation on the Sobolev classes. Submitted for publication.

von Sachs, R. & Schneider, K. (1996). Wavelet smoothing of evolutionary spectra by non-linear thresholding, *Journal of Applied and Computational Harmonic Analysis* **3**: 268–282.

Wang, Y. (1995). Jump and sharp cusp detection by wavelets, *Biometrika* **82**: 385–397.

Wang, Y. (1996). Function estimation via wavelet shrinkage for long–memory data, *Annals of Statistics* **24**: 466–484.

Young, R. K. (1993). *Wavelet Theory and its Applications*, Kluwer Academic Publishers, Boston/Dordrecht/London.

Index

Author Index

Lecture Notes in Statistics

For information about Volumes 1 to 53
please contact Springer-Verlag

Vol. 92: M. Eerola, Probabilistic Causality in Longitudinal Studies. vii, 133 pages, 1994.

Vol. 93: Bernard Van Cutsem (Editor), Classification and Dissimilarity Analysis. xiv, 238 pages, 1994.

Vol. 94: Jane F. Gentleman and G.A. Whitmore (Editors), Case Studies in Data Analysis. viii, 262 pages, 1994.

Vol. 95: Shelemyahu Zacks, Stochastic Visibility in Random Fields. x, 175 pages, 1994.

Vol. 96: Ibrahim Rahimov, Random Sums and Branching Stochastic Processes. viii, 195 pages, 1995.

Vol. 97: R. Szekli, Stochastic Ordering and Dependence in Applied Probability. viii, 194 pages, 1995.
Vol. 98: Philippe Barbe and Patrice Bertail, The Weighted Bootstrap. viii, 230 pages, 1995.

Vol. 99: C.C. Heyde (Editor), Branching Processes: Proceedings of the First World Congress. viii, 185 pages, 1995.

Vol. 100: Wlodzimierz Bryc, The Normal Distribution: Characterizations with Applications. viii, 139 pages, 1995.

Vol. 101: H.H. Andersen, M.Højbjerre, D. Sørensen, P.S.Eriksen, Linear and Graphical Models: for the Multivariate Complex Normal Distribution. x, 184 pages, 1995.

Vol. 102: A.M. Mathai, Serge B. Provost, Takesi Hayakawa, Bilinear Forms and Zonal Polynomials. x, 378 pages, 1995.

Vol. 103: Anestis Antoniadis and Georges Oppenheim (Editors), Wavelets and Statistics. vi, 411 pages, 1995.

Vol. 104: Gilg U.H. Seeber, Brian J. Francis, Reinhold Hatzinger, Gabriele Steckel-Berger (Editors), Statistical Modelling: 10th International Workshop, Innsbruck, July 10-14th 1995. x, 327 pages, 1995.

Vol. 105: Constantine Gatsonis, James S. Hodges, Robert E. Kass, Nozer D. Singpurwalla(Editors), Case Studies in Bayesian Statistics, Volume II. x, 354 pages, 1995.

Vol. 106: Harald Niederreiter, Peter Jau-Shyong Shiue (Editors), Monte Carlo and Quasi-Monte Carlo Methods in Scientific Computing. xiv, 372 pages, 1995.

Vol. 107: Masafumi Akahira, Kei Takeuchi, Non-Regular Statistical Estimation. vii, 183 pages, 1995.

Vol. 108: Wesley L. Schaible (Editor), Indirect Estimators in U.S. Federal Programs. viii, 195 pages, 1995.

Vol. 109: Helmut Rieder (Editor), Robust Statistics, Data Analysis, and Computer Intensive Methods. xiv, 427 pages, 1996.
Vol. 110: D. Bosq, Nonparametric Statistics for Stochastic Processes. xii, 169 pages, 1996.

Vol. 111: Leon Willenborg, Ton de Waal, Statistical Disclosure Control in Practice. xiv, 152 pages, 1996.

Vol. 112: Doug Fischer, Hans-J. Lenz (Editors), Learning from Data. xii, 450 pages, 1996.

Vol. 113: Rainer Schwabe, Optimum Designs for Multi-Factor Models. viii, 124 pages, 1996.

Vol. 114: C.C. Heyde, Yu. V. Prohorov, R. Pyke, and S. T. Rachev (Editors), Athens Conference on Applied Probability and Time Series Analysis Volume I: Applied Probability In Honor of J.M. Gani. viii, 424 pages, 1996.

Vol. 115: P.M. Robinson, M. Rosenblatt (Editors), Athens Conference on Applied Probability and Time Series Analysis Volume II: Time Series Analysis In Memory of E.J. Hannan. viii, 448 pages, 1996.

Vol. 116: Genshiro Kitagawa and Will Gersch, Smoothness Priors Analysis of Time Series. x, 261 pages, 1996.

Vol. 117: Paul Glasserman, Karl Sigman, David D. Yao (Editors), Stochastic Networks. xii, 298, 1996.

Vol. 118: Radford M. Neal, Bayesian Learning for Neural Networks. xv, 183, 1996.

Vol. 119: Masanao Aoki, Arthur M. Havenner, Applications of Computer Aided Time Series Modeling. ix, 329 pages, 1997.

Vol. 120: Maia Berkane, Latent Variable Modeling and Applications to Causality. vi, 288 pages, 1997.

Vol. 121: Constantine Gatsonis, James S. Hodges, Robert E. Kass, Robert McCulloch, Peter Rossi, Nozer D. Singpurwalla (Editors), Case Studies in Bayesian Statistics, Volume III. xvi, 487 pages, 1997.

Vol. 122: Timothy G. Gregoire, David R. Brillinger, Peter J. Diggle, Estelle Russek-Cohen, William G. Warren, Russell D. Wolfinger (Editors), Modeling Longitudinal and Spatially Correlated Data. x, 402 pages, 1997.

Vol. 123: D. Y. Lin and T. R. Fleming (Editors), Proceedings of the First Seattle Symposium in Biostatistics: Survival Analysis. xiii, 308 pages, 1997.

Vol. 124: Christine H. Müller, Robust Planning and Analysis of Experiments. x, 234 pages, 1997.

Vol. 125: Valerii V. Fedorov and Peter Hackl, Model-oriented Design of Experiments. viii, 117 pages, 1997.

Vol. 126: Geert Verbeke and Geert Molenberghs, Linear Mixed Models in Practice: A SAS-Oriented Approach. xiii, 306 pages, 1997.

Vol. 127: Harald Niederreiter, Peter Hellekalek, Gerhard Larcher, and Peter Zinterhof (Editors), Monte Carlo and Quasi-Monte Carlo Methods 1996, xii, 448 pp., 1997.

Vol. 128: L. Accardi and C.C. Heyde (Editors), Probability Towards 2000, x, 356 pp., 1998.

Vol. 129: Wolfgang Härdle, Gerard Kerkyacharian, Dominique Picard, and Alexander Tsybakov, Wavelets, Approximation, and Statistical Applications, xvi, 265 pp., 1998.